Universitext

Universitext

North American Editors: J.H. Ewing, F.W. Gehring, and P.R. Halmos

Berger: Geometry I, II (two volumes)
Bliedtner/Hansen: Potential Theory
Booss/Bleecker: Topology and Analysis
Chandrasekharan: Classical Fourier Transforms
Charlap: Bieberbach Groups and Flat Manifolds
Chern: Complex Manifolds Without Potential Theory
Cohn: A Classical Invitation to Algebraic Numbers and Class Fields
Curtis: Abstract Linear Algebra
Curtis: Matrix Groups
van Dalen: Logic and Structure
Devlin: Fundamentals of Contemporary Set Theory
Edwards: A Formal Background to Mathematics I a/b
Edwards: A Formal Background to Mathematics II a/b
Emery: Stochastic Calculus
Fukhs/Rokhlin: Beginner's Course in Topology
Frauenthal: Mathematical Modeling in Epidemiology
Gardiner: A First Course in Group Theory
Gårding/Tambour: Algebra for Computer Science
Godbillon: Dynamical Systems on Surfaces
Goldblatt: Orthogonality and Spacetime Geometry
Humi/Miller: Second Course in Order Ordinary Differential Equations
Hurwitz/Kritikos: Lectures on Number Theory
Iverson: Cohomology of Sheaves
Kelly/Matthews: The Non-Euclidean Hyperbolic Plane
Kostrikin: Introduction to Algebra
Krasnoselskii/Pekrovskii: Systems with Hysteresis
Luecking/Rubel: Complex Analysis: A Functional Analysis Approach
Marcus: Number Fields
McCarthy: Introduction to Arithmetical Functions
Meyer: Essential Mathematics for Applied Fields
Mines/Richman/Ruitenburg: A Course in Constructive Algebra
Moise: Introductory Problem Course in Analysis and Topology
Montesinos: Classical Tesselations and Three Manifolds
Nikulin/Shafarevich: Geometries and Groups
Øksendal: Stochastic Differential Equations
Rees: Notes on Geometry
Reisel: Elementary Theory of Metric Spaces
Rey: Introduction to Robust and Quasi-Robust Statistical Methods
Rickart: Natural Function Algebras
Rotman: Galois Theory

(continued after index)

Hans Samelson

Notes on Lie Algebras

Springer-Verlag
New York Berlin Heidelberg
London Paris Tokyo Hong Kong

Hans Samelson
Department of Mathematics
Stanford University
Stanford, CA 94305
USA

With 5 Illustrations.

First Edition of *Notes on Lie Algebras* was published in 1969 by Van Nostrand.

Mathematics Subject Classification (1980): 17 B 05-40

Library of Congress Cataloging-in-Publication Data
Samelson, Hans
Notes on Lie algebras / Hans Samelson.
p. cm.—(Universitext)
Includes bibliographical references.
ISBN-13: 978-0-387-97264-0
1. Lie algebras. I. Title.
QA252.3.S26 1990
512'.55—dc20 90-32353

Printed on acid-free paper.

Photocomposed copy prepared from the author's LaTEX file.

9 8 7 6 5 4 3 2 1

ISBN-13: 978-0-387-97264-0 e-ISBN: 978-1-4613-9014-5
DOI: 10.1007/978-1-4613-9014-5

To Nancy

time I have gone over the material in lectures at Stanford University and at the University of Crete (whose Department of Mathematics I thank for its hospitality in 1988).

The purpose, as before, is to present a simple straightforward introduction, for the general mathematical reader, to the theory of Lie algebras, specifically to the structure and the (finite dimensional) representations of the semisimple Lie algebras. I hope the book will also enable the reader to enter into the more advanced phases of the theory.

I have tried to make all arguments as simple and direct as I could, without entering into too many possible ramifications. In particular I use only the reals and the complex numbers as base fields.

The material, most of it discovered by W. Killing, E. Cartan and H. Weyl, is quite classical by now. The approach to it has changed over the years, mainly by becoming more algebraic. (In particular, the existence and the complete reducibility of representations was originally proved by Analysis; after a while algebraic proofs were found.) — The background needed for these notes is mostly linear algebra (of the geometric kind; vector spaces and linear transformations in preference to column vectors and matrices, although the latter are used too). Relevant facts and the notation are collected in the Appendix. Some familiarity with the usual general facts about groups, rings, and homomorphisms, and the standard basic facts from analysis is also assumed.

The first chapter contains the necessary general facts about Lie algebras. Semisimplicity is defined and Cartan's criterion for it in terms of a certain quadratic form, the Killing form, is developed. The chapter also brings the representations of $\mathfrak{sl}(2, \mathbb{C})$, the Lie algebra consisting of the 2×2 complex matrices with trace 0 (or, equivalently, the representations of the Lie group $SU(2)$, the 2×2 special-unitary matrices M, i.e. with $M \cdot M^* = id$ and $detM = 1$). This Lie algebra is a quite fundamental object, that crops up at many places, and thus its representations are interesting in themselves; in addition these results are used quite heavily within the theory of semisimple Lie algebras.

The second chapter brings the structure of the semisimple Lie algebras

(Cartan sub Lie algebra, roots, Weyl group, Dynkin diagram,...) and the classification, as found by Killing and Cartan (the list of all semisimple Lie algebras consists of (1) the *special- linear* ones, i.e. all matrices (of any fixed dimension) with trace 0, (2) the *orthogonal* ones, i.e. all skewsymmetric matrices (of any fixed dimension), (3) the *symplectic* ones, i.e. all matrices M (of any fixed even dimension) that satisfy $MJ = -JMT$ with a certain non-degenerate skewsymmetric matrix J, and (4) five special Lie algebras G_2, F_4, E_6, E_7, E_8, of dimensions $14, 52, 78, 133, 248$, the "exceptional Lie algebras", that just somehow appear in the process). There is also a discussion of the compact form and other real forms of a (complex) semisimple Lie algebra, and a section on automorphisms. The third chapter brings the theory of the finite dimensional representations of a semisimple Lie algebra, with the highest or extreme weight as central notion. The proof for the existence of representations is an ad hoc version of the present standard proof, but avoids explicit use of the Poincaré-Birkhoff-Witt theorem.

Complete reducibility is proved, as usual, with J.H.C. Whitehead's proof (the first proof, by H. Weyl, was analytical-topological and used the existence of a compact form of the group in question). Then come H. Weyl's formula for the character of an irreducible representation, and its consequences (the formula for the dimension of the representation, Kostant's formula for the multiplicities of the weights and algorithms for finding the weights, Steinberg's formula for the multiplicities in the splitting of a tensor product and algorithms for finding them). The last topic is the determination of which representations can be brought into orthogonal or symplectic form. This is due to I.A. Malcev; we bring the much simpler approach by Bose-Patera.

Some of the text has been rewritten and, I hope, made clearer. Errors have been eliminated; I hope no new ones have crept in. Some new material has been added, mainly the section on automorphisms, the formulas of Freudenthal and Klimyk for the multiplicities of weights, R. Brauer's algorithm for the splitting of tensor products, and the Bose-Patera proof mentioned above. The References at the end of the text contain a somewhat expanded list of books and original contributions.

In the text I use "iff" for "if and only if", "wr to" for "with respect to" and "resp." for "respectively". A reference such as "Theorem A" indicates Theorem A in the same section; a reference §m.n indicates section n in chapter m; and Ch.m refers to chapter m. The symbol [n] indicates item n in the References. The symbol "$\surd$" indicates the end of a proof, argument or discussion.

I thank Elizabeth Harvey for typing and TEXing and for support in my effort to learn TEX, and I thank Jim Milgram for help with PicTeXing the diagrams.

Hans Samelson, Stanford, September 1989

Preface to the Old Edition

These notes are a slightly expanded version of lectures given at the University of Michigan and Stanford University. Their subject, the basic facts about structure and representations of semisimple Lie algebras, due mainly to S. Lie, W. Killing, E. Cartan, and H. Weyl, is quite classical. My aim has been to follow as direct a path to these topics as I could, avoiding detours and side trips, and to keep all arguments as simple as possible. As an example, by refining a construction of Jacobson's, I get along without the enveloping algebra of a Lie algebra. (This is not to say that the enveloping algebra is not an interesting concept; in fact, for a more advanced development one certainly needs it.)

The necessary background that one should have to read these notes consists of a reasonable firm hold on linear algebra (Jordan form, spectral theorem, duality, bilinear forms, tensor products, exterior algebra,...) and the basic notions of algebra (group, ring, homomorphism,..., the Noether isomorphism theorems, the Jordan-Hoelder theorem,...), plus some notions of calculus. The principal notions of linear algebra used are collected, not very systematically, in an appendix; it might be well for the reader to glance at the appendix to begin with, if only to get acquainted with some of the notation. I restrict myself to the standard fields: $\mathbb{R}$ = reals, $\mathbb{C}$ = complex numbers ($\bar{a}$ denotes the complex-conjugate of a); $\mathbb{Z}$ denotes the integers; $\mathbb{Z}_n$ is the cyclic group of order n. "iff" means "if and only if"; "w.r.to" means "with respect to". In the preparation of these notes, I substituted my own version of the Halmos-symbol that indicates the end of a proof or an argument; I use "$\surd$". The bibliography is kept to a minimum; Jacobson's book contains a fairly extensive list of references and some historical comments. Besides the standard sources I have made use of mimeographed notes that I have come across (Albert, van Est, Freudenthal, Mostow, J. Shoenfield).

Stanford, 1969

Contents

1

Generalities

1.1 Basic definitions, examples

A *multiplication* or *product* on a vector space V is a bilinear map from $V \times V$ to V.

Now comes the definition of the central notion of this book:

A *Lie algebra* consists of a (finite dimensional) vector space, over a field $\mathbb{F}$, and a multiplication on the vector space (denoted by [ ], pronounced "bracket", the image of a pair (X, Y) of vectors denoted by $[XY]$ or $[X, Y]$), with the properties

(a) $\quad [XX] = 0,$

(b) $\quad [X[YZ]] + [Y[ZX]] + [Z[XY]] = 0$

for all elements X, resp X, Y, Z, of our vector space.

Property (a) is called skew-symmetry; because of bilinearity it implies (and is implied by, if the characteristic of $\mathbb{F}$ is not 2)

(a′) $\quad [XY] = -[YX].$

(For $\Rightarrow$ replace X by $X + Y$ in (a) and expand by bilinearity; for $\Leftarrow$ put $X = Y$ in (a), getting $2[XX] = 0$.)

In more abstract terms (a) says that [ ] is a linear map from the second exterior power of the vector space to the vector space.

Property (b) is called the *Jacobi identity*; it is related to the usual associative law, as the examples will show.

Usually we denote Lie algebras by small German letters: $\mathfrak{a}, \mathfrak{b}, \ldots, \mathfrak{g}, \ldots$.

Naturally one could generalize the definition, by allowing the vector space to be of infinite dimension or by replacing "vector space" by "module over a ring".

Note: From here on we use for $\mathbb{F}$ only the reals, $\mathbb{R}$, or the complexes, $\mathbb{C}$. Some of the following examples make sense for any field $\mathbb{F}$.

Example 0: Any vector space with $[XY] = 0$ for all X, Y; these are the *Abelian* Lie algebras.

Example 1: Let A be an algebra over $\mathbb{F}$ (a vector space with an associative multiplication $X \cdot Y$). We make A into a Lie algebra A_L (also called *A as Lie algebra)* by defining $[XY] = X \cdot Y - Y \cdot X$. The Jacobi identity holds; just "multiply out".

As a simple case, $\mathbb{F}_L$ is the *trivial* Lie algebra, of dimension 1 and Abelian. For another "concrete" case see Example 12.

Example 2: A special case of Example 1: Take for A the algebra of all operators (endomorphisms) of a vector space V; the corresponding A_L is called the *general Lie algebra of* V, $\mathfrak{gl}(V)$. Concretely, taking number space $\mathbb{R}^n$ as V, this is the *general linear Lie algebra* $\mathfrak{gl}(n, \mathbb{R})$ of all $n \times n$ real matrices, with $[XY] = XY - YX$. Similarly $\mathfrak{gl}(n, \mathbb{C})$.

Example 3: The *special linear Lie algebra* $\mathfrak{sl}(n, \mathbb{R})$ consists of all $n \times n$ real matrices with trace 0 (and has the same linear and bracket operations as $\mathfrak{gl}(n, \mathbb{R})$—it is a "sub Lie algebra"); similarly for $\mathbb{C}$. For any vector space V we have $\mathfrak{sl}(V)$, the *special linear Lie algebra of* V, consisting of the operators on V of trace 0.

Example 4: Let V be a vector space, and let b be a non-degenerate symmetric bilinear form on V. The *orthogonal Lie algebra* $\mathfrak{o}(V, b)$, or just $\mathfrak{o}(V)$ if it is clear which b is intended, consists of all operators T on V under which the form b is "infinitesimally invariant" (see §1.3 for explanation of the term), i.e., that satisfy $b(Tv, w) + b(v, Tw) = 0$ for all v, w in V, or equivalently $b(Tv, v) = 0$ for all v in V; again the linear and bracket operations are as in $\mathfrak{gl}(V)$. One has to check of course that $[ST]$ leaves b infinitesimally invariant, if S and T do; this is elementary.

For $V = \mathbb{F}^n$ one usually takes for $b(X, Y)$ the form $\Sigma x_i y_i = X^\top \cdot Y$ with $X = (x_1, x_2, \ldots, x_n)$, $Y = (y_1, y_2, \ldots, y_n)$; one writes $\mathfrak{o}(n, \mathbb{F})$ for the corresponding orthogonal Lie algebra. The infinitesimal invariance property reads now $X^\top(M^\top + M)Y = 0$ and so $\mathfrak{o}(n, \mathbb{F})$ consists of the matrices M over $\mathbb{F}$ that satisfy $M^\top + M = 0$, i.e., the skew-symmetric ones. $\mathbb{F} = \mathbb{R}$ is the standard case; but the case $\mathbb{C}$ (complex skew matrices) is also important.

Example 5: Let V be a complex vector space, and let c be a Hermitean (positive definite) inner product on V. The *unitary Lie algebra* $\mathfrak{u}(V, c)$, or just $\mathfrak{u}(V)$, consists of the operators T on V with the infinitesimal invariance property $c(TX, Y) + c(X, TY) = 0$. This is a Lie algebra over $\mathbb{R}$, but not over $\mathbb{C}$ (if T has the invariance property, so does rT for real r, but not iT—because c is conjugate-linear in the first variable—unless T is 0).

For $V = \mathbb{C}^n$ and $c(X, Y) = \Sigma \bar{x}_i \cdot y_i$ (the "$^-$" meaning complex-conjugate) this gives the Lie algebra $\mathfrak{u}(n)$, consisting of the matrices M that satisfy $M^* + M = 0$ (where * means *transpose conjugate* or *adjoint*), i.e., the *skew-Hermitean* ones.

There is also the *special unitary Lie algebra* $\mathfrak{su}(V)$ (or $\mathfrak{su}(n)$), consisting of the elements of $\mathfrak{u}(V)$ (or $\mathfrak{u}(n)$) of trace 0.

Example 6: Let V be a vector space over $\mathbb{F}$, and let Ω be a non-degenerate skew-symmetric bilinear form on V. The *symplectic Lie algebra* $\mathfrak{sp}(V, \Omega)$ or just $\mathfrak{sp}(V)$ consists of the operators T on V that leave Ω infinitesimally invariant: $\Omega(TX, Y) + \Omega(X, TY) = 0$.

One writes $\mathfrak{sp}(n, \mathbb{R})$ and $\mathfrak{sp}(n, \mathbb{C})$ for the symplectic Lie algebras of $\mathbb{R}^{2n}$ and $\mathbb{C}^{2n}$ with $\Omega(X, Y) = x_1y_2 - x_2y_1 + x_3y_4 - x_4y_3 + \cdots + x_{2n-1}y_{2n} - x_{2n}y_{2n-1}$. (It is well known that non-degeneracy of Ω requires $\dim V$ even and that Ω has the form just shown wr to a suitable coordinate system.)

With $J_1 = \begin{bmatrix} 0 & 1 \\ -1 & 0 \end{bmatrix}$ and $J = \mathrm{diag}(J_1, J_1, \ldots, J_1)$ this can also be described as the set of $2n \times 2n$ matrices that satisfy $M^\top J + JM = 0$.

The matrices simultaneously in $\mathfrak{sp}(n, \mathbb{C})$ and in $\mathfrak{u}(2n)$ form a real Lie algebra, denoted by $\mathfrak{sp}(n)$. (An invariant definition for $\mathfrak{sp}(n)$ is as follows: Let c and Ω be defined as in Examples 5 and 6, on the same vector space V, of dimension $2n$. They define, respectively, a conjugate-linear map C and a linear map L of V to its dual space $V^\top$. Then $J = L^{-1} \cdot C$ is a conjugate-linear map of V to itself. If $J^2 = -id$, then (c, Ω) is called a symplectic pair, and in that case the symplectic Lie algebra $\mathfrak{sp}(c, \Omega)$ is defined as the intersection $\mathfrak{u}(c) \cap \mathfrak{sp}(\Omega)$.)

We introduce the classical, standard, symbols for these Lie algebras: $\mathfrak{sl}(n+1, \mathbb{C})$ is denoted by A_n, for $n = 1, 2, 3, \ldots$; $\mathfrak{o}(2n+1, \mathbb{C})$, for $n = 2, 3, 4, \ldots$, is denoted by B_n; $\mathfrak{sp}(n, \mathbb{C})$, for $n = 3, 4, 5, \ldots$, is denoted by C_n; finally $\mathfrak{o}(2n, \mathbb{C})$, for $n = 4, 5, 6, \ldots$, is denoted by D_n.(We shall use these symbols, in deviation from our convention on notation for Lie algebras.) The same symbols are used for the case $\mathbb{F} = \mathbb{R}$.

The A_l, B_l, C_l, D_l are the *four families* of the *classical* Lie algebras. The restrictions on n are made to prevent "double exposure": one has the (not quite obvious) relations $B_1 \approx C_1 \approx A_1; C_2 \approx B_2; D_3 \approx A_3; D_2 \approx A_1 \oplus A_1; D_1$ is Abelian of dimension 1. (See §1.4 for $\approx$ and $\oplus$.)

Example 7: We describe the orthogonal Lie algebra $\mathfrak{o}(3)$ in more detail. Let R_x, R_y, R_z denote the three matrices

$$\begin{bmatrix} 0 & 0 & 0 \\ 0 & 0 & -1 \\ 0 & 1 & 0 \end{bmatrix}, \quad \begin{bmatrix} 0 & 0 & 1 \\ 0 & 0 & 0 \\ -1 & 0 & 0 \end{bmatrix}, \quad \begin{bmatrix} 0 & -1 & 0 \\ 1 & 0 & 0 \\ 0 & 0 & 0 \end{bmatrix}$$

(These are the "infinitesimal rotations" around the x- or y- or z-axis, see §1.3.) Clearly they are a basis for $\mathfrak{o}(3)$ (3×3 real skew matrices); they are also a basis, over $\mathbb{C}$, for $\mathfrak{o}(3, \mathbb{C})$. One computes

$$[R_x R_y] = R_z \ , \quad [R_y R_z] = R_x \ , \quad [R_z R_x] = R_y \ .$$

Example 8: $\mathfrak{su}(2)$ in detail (2×2 skew-Hermitean, trace 0). The following three matrices S_x, S_y, S_z clearly form a basis (about the reasons for choosing these particular matrices see §1.4):

$$1/2\begin{bmatrix}0 & i\\ i & 0\end{bmatrix}, \quad 1/2\begin{bmatrix}0 & -1\\ 1 & 0\end{bmatrix}, \quad 1/2\begin{bmatrix}i & 0\\ 0 & -i\end{bmatrix}$$

One verifies $[S_xS_y] = S_z$, $[S_yS_z] = S_x$, $[S_zS_x] = S_y$. Note the similarity to Example 7, an example of an isomorphism, cf. §1.4.

Example 9: The Lie algebra $\mathfrak{sl}(2,\mathbb{C})$ (or A_1), 2×2 matrices of trace 0. A basis is given by the three matrices

$$H = \begin{bmatrix}1 & 0\\ 0 & -1\end{bmatrix}, \quad X_+ = \begin{bmatrix}0 & 1\\ 0 & 0\end{bmatrix}, \quad X_- = \begin{bmatrix}0 & 0\\ 1 & 0\end{bmatrix}$$

One computes $[HX_+] = 2X_+$, $[HX_-] = -2X_-$, $[X_+X_-] = H$. This Lie algebra and these relations will play a considerable role later on.

The standard skew-symmetric (exterior) form $\det[X,Y] = x_1y_2 - x_2y_1$ on $\mathbb{C}^2$ is invariant under $\mathfrak{sl}(2,\mathbb{C})$ (precisely because of the vanishing of the trace), and so $\mathfrak{sl}(2,\mathbb{C})$ is identical with $\mathfrak{sp}(1,\mathbb{C})$. Thus $A_1 = C_1$.

Example 10: The *affine* Lie algebra *of the line*, $\mathfrak{aff}(1)$. It consists of all real 2×2 matrices with second row 0. The two elements

$$X_1 = \begin{bmatrix}1 & 0\\ 0 & 0\end{bmatrix}, \quad X_2 = \begin{bmatrix}0 & 1\\ 0 & 0\end{bmatrix}$$

form a basis, and we have $[X_1X_2] = X_2$. (See "affine group of the line", §1.3.)

Example 11: The *Lorentz* Lie algebra $\mathfrak{o}(3,1;\mathbb{R})$, or $\mathfrak{l}_{3,1}$ in short (corresponding to the well known Lorentz group of relativity). In $\mathbb{R}^4$, with vectors written as $v = (x,y,z,t)$, we use the *Lorentz inner product* $\langle v,v\rangle_L = x^2+y^2+z^2-t^2$; putting $I_{3,1} = \mathrm{diag}(1,1,1,-1)$ and considering v as column vector, this is also $v^\top I_{3,1}v$. Now $\mathfrak{l}_{3,1}$ consists of those operators T on $\mathbb{R}^4$ that leave $\langle\cdot,\cdot\rangle_L$ infinitesimally invariant (i.e., $\langle Tv,w\rangle_L + \langle v,Tw\rangle_L = 0$ for all v,w), or of the 4×4 real matrices M with $M^\top I_{3,1} + I_{3,1}M = 0$.

Example 12: We consider the algebra $\mathbb{H}$ of the *quaternions*, over $\mathbb{R}$, with the usual basis $1,i,j,k$; 1 is unit, $i^2 = j^2 = k^2 = -1$ and $ij = -ji = k$,etc. Any quaternion can be written uniquely in the form $a + jb$ with a,b in $\mathbb{C}$. Associating with this quaternion the matrix

$$\begin{bmatrix}a & -\bar{b}\\ b & \bar{a}\end{bmatrix}$$

sets up an isomorphism of the quaternions with the $\mathbb{R}$–algebra of 2×2 complex matrices of this form.

Such a matrix in turn can be written in the form $rI + M$ with real r and M skew-Hermitean with trace 0. This means that the quaternions as Lie algebra are isomorphic (see §1.4) to the direct sum (see §1.4 again) of the Lie algebras $\mathbb{R}$ (i.e.,$\mathbb{R}_L$) and $\mathfrak{su}(2)$(Example 8).

1.2 Structure constants

Let $\mathfrak{g}$ be a Lie algebra and take a basis $\{X_1, X_2, \ldots, X_n\}$ for (the vector space) $\mathfrak{g}$. By bilinearity the [ ]-operation in $\mathfrak{g}$ is completely determined once the values $[X_iX_j]$ are known. We "know" them by writing them as linear combinations of the X_i. The coefficients c_{ij}^k in the relations $[X_iX_j] = c_{ij}^k X_k$ (sum over repeated indices!) are called the *structure constants* of $\mathfrak{g}$ (relative to the given basis). [Examples 7–10 are of this kind; e.g., in Example 10 we have $c_{12}^1 = 0$, $c_{12}^2 = 1$; for $i = j$ one gets 0 of course.] Axioms (a) and (b) of §1.1 find their expressions in the relations $c_{ij}^k = -c_{ji}^k$ ($= 0$, if $i = j$) and $c_{il}^m c_{jk}^l + c_{jl}^m c_{ki}^l + c_{kl}^m c_{ij}^l = 0$. Under change of basis the structure constants change as a tensor of type $(2,1)$: if $X_j' = a_j^i X_i$, then $c'^k_{ij} \cdot a_k^l = c_{rs}^l \cdot a_i^r \cdot a_j^s$.

We interpret this as follows: Let $\dim \mathfrak{g} = n$, and let $\mathbb{F}$ be the field under consideration. We consider the n^3-dimensional vector space of systems c_{ij}^k, with $i, j, k = 1, \ldots, n$. The systems that form the structure constants of some Lie algebra form an algebraic set S, defined by the above linear and quadratic equations that correspond to axioms (a) and (b) of §1.1. The *general linear group* $GL(n, \mathbb{F})$, which consists of all invertible $n \times n$ matrices over $\mathbb{F}$, operates on S, by the formulae above. The various systems of structure constants of a given Lie algebra relative to all its bases form an *orbit* (set of all transforms of one element) under this action. Conversely, the systems of structure constants in an orbit can be interpreted as giving rise to one and the same Lie algebra. Thus there is a natural bijection between orbits (of systems of structure constants) and isomorphism classes of Lie algebras (of dimension n); see §1.4 for "isomorphism". As an example, the orbit of the system "$c_{ij}^k = 0$ for all i, j, k", which clearly consists of just that one system, corresponds to "the" Lie algebra (of dimension n) with $[XY] = 0$ for all X, Y, i.e., "the" Abelian Lie algebra of $\dim n$.

1.3 Relations with Lie groups

We discuss only the beginning of this topic. First we look at the Lie groups corresponding to the Lie algebras considered in §1.1.

The *general linear group* $GL(n, \mathbb{F})$ consists of all invertible $n \times n$ matrices over $\mathbb{F}$.

The *special linear group* $SL(n, \mathbb{F})$ consists of the elements of $GL(n, \mathbb{F})$ with determinant 1.

The (real) *orthogonal group* $O(n, \mathbb{R})$ or just $O(n)$ consists of the real $n \times n$ matrices M with $M^\top \cdot M = 1$; for the *complex orthogonal group* $O(n, \mathbb{C})$ we replace "real" by "complex" in the definition. The *special (real) orthogonal group* $SO(n, \mathbb{R}) = SO(n)$ is $O(n) \cap SL(n, \mathbb{R})$; similarly for $SO(n, \mathbb{C})$.

The *unitary group* $U(n)$ consists of all the (complex) matrices M with $M^* \cdot M = 1$; the *special unitary group* $SU(n)$ is $U(n) \cap SL(n, \mathbb{C})$.

The *symplectic group* $Sp(n, \mathbb{F})$ consists of all $2n \times 2n$ matrices over $\mathbb{F}$ with $M^\top \cdot J \cdot M = J$ (see §1.2 for J); such matrices automatically have det $= 1$ (best proved by considering the element Ω^n in the exterior algebra, with the Ω of §1.2). The *symplectic group* $Sp(n)$ is $Sp(n, \mathbb{C}) \cap U(2n)$. (All these definitions can be made invariantly, as in §1.2 for Lie algebras.)

The *affine group of the line*, $Aff(1)$, consists of all real, 2×2, invertible matrices with second row $(0, 1)$, i.e., the transformations $x' = ax + b$ of the real line with $a \neq 0$.

Finally the *Lorentz group* consists of all real 4×4 matrices M with $M^\top I_{3,1} M = I_{3,1}$.

The set of all $n \times n$ matrices over $\mathbb{F}$ has an obvious identification with the standard vector space of dimension n^2 over $\mathbb{F}$. Thus all the groups defined above are subsets of various spaces $\mathbb{R}^m$ or $\mathbb{C}^m$, defined by a finite number of simple equations (like the relations $M^\top \cdot M = I$ for $O(n, \mathbb{F})$). In fact, they are algebraic varieties (except for $U(n)$ and $SU(n)$, where the presence of complex conjugation interferes slightly). It is fairly obvious that they are all topological manifolds, in fact differentiable, infinitely differentiable, real-analytic, and some of them even complex holomorphic. (Also $O(n)$, $SO(n)$, $U(n)$, $SU(n)$, $Sp(n)$ are easily seen to be compact, namely closed and bounded in their respective spaces.)

We now come to the relation of these groups with the corresponding Lie algebras.

Briefly, a Lie algebra is the tangent space of a Lie group at the unit element.

For $\mathfrak{gl}(n, \mathbb{F})$ we take a smooth curve $M(t)$ in $GL(n, \mathbb{F})$ (so each $M(t)$ is an invertible matrix over $\mathbb{F}$) with $M(0) = I$. The tangent vector at $t = 0$, i.e., the derivative $M'(0)$, is then an element of $\mathfrak{gl}(n, \mathbb{F})$. Every element of $\mathfrak{gl}(n, \mathbb{F})$ appears for a suitably chosen curve. It is worthwhile to point out a special way of producing these curves:

Given an element X of $\mathfrak{gl}(n, \mathbb{F})$, with $\mathbb{F} = \mathbb{R}$ or $\mathbb{C}$, i.e., an $n \times n$ matrix, we take a variable s in $\mathbb{F}$ and form $e^{sX} = \Sigma s^i X^i / i!$ (also written as $\exp(sX)$; this series of matrices is as well behaved as the usual exponential function. For each value of s it gives an invertible matrix, i.e., one in $GL(n, \mathbb{F})$; one has $\exp(0X) = \exp(0) = I$ and $e^{sX} \cdot e^{s'X} = e^{(s+s')X}$. Thus the curve $\exp(sX)$, with s running over $\mathbb{R}$, is a group, called the *one-parameter group* determined by X. (Strictly speaking the one-parameter group is the *map* that sends s to $\exp(sX)$.) We get X back from the one-parameter group by taking the derivative wr to s for $s = 0$.

For $O(n, \mathbb{F})$ we take a curve consisting of orthogonal matrices, so that $M^\top(t) \cdot M(t) = I$ for all t. Differentiating and putting $t = 0$, we find $(M'(0))^\top + M'(0) = 0$ (remember $M(0) = I$); so our $X = M'(0)$ lies in $\mathfrak{o}(n, \mathbb{F})$. Conversely, take X with $X^\top + X = 0$; form $\exp(sX^\top) \cdot \exp(sX)$ and differentiate it. The result can be written as $\exp(sX^\top) \cdot X^\top \cdot \exp(sX) + \exp(sX^\top) \cdot X \cdot \exp(sX)$, which on account of $X^\top + X = 0$ is identically 0. Thus $\exp(sX^\top) \cdot \exp(sX)$ is constant; taking $s = 0$, we see that the constant is I, meaning that $\exp(sX)$ lies in $O(n, \mathbb{F})$ for all s.

Similar considerations hold for the other groups. In particular, X has trace 0 (i.e., belongs to $\mathfrak{sl}(n, \mathbb{F})$), iff $\det \exp(sX) = 1$ for all s (because of $\det \exp X = \exp(\mathrm{tr} X)$). X is skew-Hermitean (belongs to $\mathfrak{u}(n)$), iff all $\exp(sX)$ are unitary. X satisfies $X^\top \cdot J + J \cdot X = 0$ (it belongs to $\mathfrak{sp}(n, \mathbb{F})$), iff the relation $\exp(sX^\top) \cdot J \cdot \exp(sX) = J$ holds for all s (all the $\exp(sX)$ belong to $Sp(n, \mathbb{F})$). Etc.

As for the "infinitesimal invariance" of §1.2, it is simply the infinitesimal form of the relation that defines $O(n, \mathbb{F})$: With the form b of §1.1, Example 4, we let $g(t)$ be a smooth one-parameter family of isometries of V, so that $b(g(t)v, g(t)w) = b(v, w)$ for all t, with $g(0) = id$. Taking the derivative for $t = 0$ and putting $g'(0) = T$, we get $b(Tv, w) + b(v, Tw) = 0$. (As we saw above, in matrix language this says $X^\top + X = 0$.)—Similarly for the other examples.

This is a good point to indicate some reasons why, for X, Y in $\mathfrak{gl}(n, \mathbb{F})$, the combination $[XY] = XY - YX$ is important:

(1) Put $f(s) = \exp(sX) \cdot Y \cdot \exp(-sX)$; i.e., form the conjugate of Y by $\exp(sX)$. The derivative of f for $s = 0$ is then $XY - YX$ (and the Taylor expansion of f is $f(s) = Y + s[XY] + \ldots$).

(2) Let $g(s)$ be the commutator $\exp(sX) \cdot \exp(sY) \cdot \exp(-sX) \cdot \exp(-sY)$. One finds $g(0) = I$, $g'(0) = 0$, $g''(0) = 2(XY - YX) = 2[XY]$; the Taylor expansion is $g(s) = I + s^2[XY] + \ldots$

In both cases we see that $[XY]$ is some measure of non-commutativity.

1.4 Elementary algebraic concepts

Let $\mathfrak{g}$ be a Lie algebra. For two subspaces A, B of $\mathfrak{g}$ the symbol $[AB]$ denotes the linear span of the set of all $[XY]$ with X in A and Y in B; occasionally this notation is also used for arbitrary subsets A, B. Similarly, and more elementary, one defines $A + B$.

A *sub Lie algebra* of $\mathfrak{g}$ is a subspace, say $\mathfrak{q}$, of $\mathfrak{g}$ that is closed under the bracket operation (i.e., $[\mathfrak{q}\mathfrak{q}] \subset \mathfrak{q}$); $\mathfrak{q}$ becomes then a Lie algebra with the linear and bracket operations inherited from $\mathfrak{g}$. (Examples #3–6 in §1.1 are sub Lie algebras of the relevant general linear Lie algebras.)

A sub Lie algebra $\mathfrak{q}$ is an *ideal* of $\mathfrak{g}$ if $[\mathfrak{g}\mathfrak{q}] \subset \mathfrak{q}$ (if $X \in \mathfrak{g}$ and $Y \in \mathfrak{q}$ implies $[XY] \in \mathfrak{q}$). By skew-symmetry (property (a) in §1.1) ideals are automatically two-sided: $[\mathfrak{g}\mathfrak{q}] = [\mathfrak{q}\mathfrak{g}]$. If $\mathfrak{q}$ is an ideal, then the quotient space $\mathfrak{g}/\mathfrak{q}$ (whose elements are the linear cosets $X + \mathfrak{q}$) carries an induced [ ]-operation, defined by $[X+\mathfrak{q}, Y+\mathfrak{q}] = [XY]+\mathfrak{q}$; as in ordinary algebra one verifies that this is well defined, i.e., does not depend on the choice of the representatives X, Y. With this operation $\mathfrak{g}/\mathfrak{q}$ becomes a Lie algebra, the *quotient Lie algebra* of $\mathfrak{g}$ by $\mathfrak{q}$. For a trivial example: every subspace of an Abelian Lie algebra is an ideal.

A *homomorphism*, say φ, from a Lie algebra $\mathfrak{g}$ to a Lie algebra $\mathfrak{g}_1$ is a linear map $\varphi : \mathfrak{g} \to \mathfrak{g}_1$ that preserves brackets: $\varphi([XY]) = [\varphi(X), \varphi(Y)]$. (If $\mathfrak{g} = \mathfrak{g}_1$, we speak of an *endomorphism*.) A homomorphism is an *isomorphism* (symbol $\approx$), if it is one in the sense of linear maps, i.e., if it is injective and surjective; the inverse map is then also an isomorphism of Lie algebras.

Implicitly we used the concept "isomorphism" already in §1.2, when we acted as if a Lie algebra were determined by its structure constants (wr to some basis), e.g., when we talked about "the" Abelian Lie algebra of dimension n; what we meant was of course "determined up to isomorphism".

An isomorphism of a Lie algebra with itself is an *automorphism*.

A not quite trivial isomorphism occurs in §1.1, Examples 6 and 7: $\mathfrak{su}(2)$ and $\mathfrak{o}(3)$ are isomorphic, via the map $S_x \to R_x$ etc. (After complexifying - see below - this is the isomorphism $A_1 \approx B_1$ mentioned in §1.2.)

It is interesting, and we explain it in more detail: Consider the group $SO(3)$ of rotations of $\mathbb{R}^3$ or, equivalently, of the 2-sphere S^2. By stereographic projection these rotations turn into fractional linear transformations of a complex variable, namely those of the form

$$z' = \frac{az + b}{-\bar{b}z + \bar{a}}$$

with $a \cdot \bar{a} + b \cdot \bar{b} = 1$. The matrices

$$\begin{bmatrix} a & b \\ -\bar{b} & \bar{a} \end{bmatrix}$$

with $|a|^2 + |b|^2 = 1$ occurring here make up exactly the group $SU(2)$. However the matrix is determined by the transformation above only up to sign; we have a double-valued map. Going in the opposite direction, we have here a homomorphism of $SU(2)$ onto $SO(3)$, whose kernel consists of I and $-I$. This is a local isomorphism, i.e., it maps a small neighborhood of I in $SU(2)$ bijectively onto a neighborhood of I in $SO(3)$. There is then an induced isomorphism of the Lie algebras (= tangent spaces at the unit elements); and that is the isomorphism from $\mathfrak{su}(2)$ to $\mathfrak{o}(3)$ above.

We take up one more example of an isomorphism, of interest in physics: The Lorentz Lie algebra $\mathfrak{l}_{3,1}$ (see Example 11 in §1.1) is isomorphic to $\mathfrak{sl}(2,\mathbb{C})_{\mathbb{R}}$ (the latter meaning $\mathfrak{sl}(2,\mathbb{C})$ considered over $\mathbb{R}$ only—the realification(see below)). Actually this is easier to understand for the corresponding groups. Let U be the 4-dimensional real vector space consisting of the 2×2 (complex) Hermitean matrices. The function det (= determinant) from U to $\mathbb{R}$ happens to be a *quadratic* function on U; and with a simple change of variables it becomes (up to a sign) equal to the Lorentz form $\langle \cdot, \cdot \rangle_L$: with $M = \begin{bmatrix} \alpha & \beta + i\gamma \\ \beta - i\gamma & \delta \end{bmatrix}$ we put $\alpha = t - x$, $\delta = t + x$, $\beta = y$, $\gamma = z$ and get $\det M = t^2 - x^2 - y^2 - z^2$. Now $SL(2,\mathbb{C})$ acts on U in a natural way, via $M \to A^*MA$ for $A \in SL(2,\mathbb{C})$ and $M \in U$. Because of the multiplicative nature of det and the given fact $\det A = 1$ we find $\det A^*MA = \det M$, i.e., A leaves the Lorentz inner product invariant, and we have here a homomorphism of $SL(2,\mathbb{C})$ into the Lorentz group. The kernel of the map is easily seen to consist of id and −id. The map is also surjective—we shall not go into details here. (Thus the relation between the two groups is similar to that between $SO(3)$ and $SU(2)$—the former is quotient of the latter by a $\mathbb{Z}/2$.) Infinitesimally this means that the Lie algebras of $SL(2,\mathbb{C})$ and the Lorentz group are isomorphic. In detail, to X in $\mathfrak{sl}(2,\mathbb{C})$ we assign the operator on U defined by $M \to X^*M + MX$ (put $A = \exp(tX)$ above and differentiate); and this operator will leave the Lorentz form (i.e., $\det M$) invariant in the infinitesimal sense (one can also verify this by an algebraic computation, based on $\mathrm{tr} X = 0$).

A *representation* of a Lie algebra $\mathfrak{g}$ on a vector space V is a homomorphism, say φ, of $\mathfrak{g}$ into the general linear algebra $\mathfrak{gl}(V)$ of V. (We allow the possibility of $\mathfrak{g}$ real, but V complex; this means that temporarily one considers $\mathfrak{gl}(V)$ as a real Lie algebra, by "restriction of scalars".) φ assigns to each X in $\mathfrak{g}$ an operator $\varphi(X) : V \to V$ (or, if one wants to use a basis of V, a matrix), depending linearly on X (so that $\varphi(aX + bY) = a\varphi(X) + b\varphi(Y)$) and satisfying $\varphi([XY]) = [\varphi(X), \varphi(Y)]$ $(= \varphi(X)\varphi(Y) - \varphi(Y)\varphi(X))$ ("preservation of brackets"). [One often writes $X \cdot v$ or $X.v$ or simply Xv instead of $\varphi(X)(v)$ (the image of the vector v under the operator $\varphi(X)$); one even talks about the operator X, meaning the operator $\varphi(X)$. Preservation of bracket appears then in the form $[XY]v = XYv - YXv$.] One says that $\mathfrak{g}$ *acts* or *operates* on V, or that V is a $\mathfrak{g}$-space (or $\mathfrak{g}$-module). Note that Examples 2–11 of §1.1 all come equipped with an obvious representation—their elements are given as operators on certain vector spaces, and $[XY]$ equals $XY - YX$ by definition. Of course these Lie algebras may very well have representations on some other vector spaces; in fact they do, and the study of these possibilities is one of our main aims.

The *kernel* of a homomorphism $\varphi : \mathfrak{g} \to \mathfrak{g}_1$ is the set $\varphi^{-1}(0)$ of all X in $\mathfrak{g}$ with φ-image 0; it is easily seen to be an ideal in $\mathfrak{g}$; we write ker φ for it. More generally, the inverse image under φ of a sub Lie algebra, resp. ideal of $\mathfrak{g}_1$, is a sub Lie algebra, resp. ideal of $\mathfrak{g}$. The *image* $\varphi(\mathfrak{g})$ (also denoted by im φ) is a sub Lie algebra of $\mathfrak{g}_1$, as is the image of any sub Lie algebra of $\mathfrak{g}$.

Conversely, if $\mathfrak{q}$ is an ideal of $\mathfrak{g}$, then the natural map π of $\mathfrak{g}$ into the quotient Lie algebra $\mathfrak{g}/\mathfrak{q}$, defined by $X \to X+\mathfrak{q}$, is a homomorphism, whose kernel is exactly $\mathfrak{q}$ and which is surjective. In other words, there is a natural "short exact sequence" $0 \to \mathfrak{q} \to \mathfrak{g} \to \mathfrak{g}/\mathfrak{q} \to 0$. If ψ is a homomorphism of $\mathfrak{g}$ into some Lie algebra $\mathfrak{g}_1$ that sends $\mathfrak{q}$ to 0, then it "factors through π": There is a (unique) homomorphism $\psi' : \mathfrak{g}/\mathfrak{q} \to \mathfrak{g}_1$ with $\psi = \psi' \circ \pi$; the formula $\psi'(X+\mathfrak{q}) = \psi(X)$ clearly gives a well-defined linear map, and from the definition of [ ] in $\mathfrak{g}/\mathfrak{q}$ it is clear that ψ' preserves [ ].

There is the *first isomorphism theorem* (analogous to that of group theory): let $\mathfrak{q}$ be the kernel of the homomorphism $\varphi : \mathfrak{g} \to \mathfrak{g}_1$; the induced map φ' sets up an isomorphism of $\mathfrak{g}/\mathfrak{q}$ with the image Lie algebra $\varphi(\mathfrak{g})$.

For the proof we note that clearly $\operatorname{im}\varphi = \operatorname{im}\varphi'$ so that the map in question is surjective; it is also injective since the only coset of $\mathfrak{q}$ with φ-image 0 is clearly $\mathfrak{q}$ itself. An easy consequence of this is the following: Let $\mathfrak{a}$ and $\mathfrak{b}$ be ideals in $\mathfrak{g}$, with $\mathfrak{a} \subset \mathfrak{b}$; then the natural maps give rise to an isomorphism $\mathfrak{g}/\mathfrak{b} \approx (\mathfrak{g}/\mathfrak{a})/(\mathfrak{b}/\mathfrak{a})$.

Next: if $\mathfrak{a}$ and $\mathfrak{b}$ are ideals of $\mathfrak{g}$, so are $\mathfrak{a}+\mathfrak{b}$ and $[\mathfrak{a}\mathfrak{b}]$; if $\mathfrak{a}$ is an ideal and $\mathfrak{b}$ a sub Lie algebra, then $\mathfrak{a}+\mathfrak{b}$ is a sub Lie algebra. The proof for $\mathfrak{a}+\mathfrak{b}$ is trivial; that for $[\mathfrak{a}\mathfrak{b}]$ uses the Jacobi identity.

The intersection of two sub Lie algebras is again a sub Lie algebra, of course; if $\mathfrak{a}$ is a sub Lie algebra and $\mathfrak{b}$ is an ideal of $\mathfrak{g}$, then $\mathfrak{a}\cap\mathfrak{b}$ is an ideal of $\mathfrak{a}$. The *second isomorphism theorem* says that in this situation the natural map of $\mathfrak{a}$ into $\mathfrak{a}+\mathfrak{b}$ induces an isomorphism of $\mathfrak{a}/\mathfrak{a}\cap\mathfrak{b}$ with $(\mathfrak{a}+\mathfrak{b})/\mathfrak{b}$; we forego the standard proof.

Two elements X and Y of $\mathfrak{g}$ are said to *commute*, if $[XY]$ is 0. (The term comes from the fact that in the case $\mathfrak{g} = \mathfrak{gl}(n, \mathbb{F})$ (or any A_L) the condition $[XY] = 0$ just means $XY = YX$; it is also equivalent to the condition that all $\exp(sX)$ commute with all $\exp(tY)$ (see §1.3 for exp.).) The *centralizer* $\mathfrak{g}_S$ of a subset S of $\mathfrak{g}$ is the set (in fact a sub Lie algebra) of those X in $\mathfrak{g}$ that commute with all Y in S. For $S = \mathfrak{g}$ this is the *center* of $\mathfrak{g}$. Similarly the *normalizer* of a sub Lie algebra $\mathfrak{a}$ consists of the X in $\mathfrak{g}$ with $[X\mathfrak{a}] \subset \mathfrak{a}$; it is a sub Lie algebra of $\mathfrak{g}$, and contains $\mathfrak{a}$ as an ideal (and is the largest sub Lie algebra of $\mathfrak{g}$ with this property).

The (external) *direct sum* of two Lie algebras $\mathfrak{g}_1, \mathfrak{g}_2$, written $\mathfrak{g}_1 \oplus \mathfrak{g}_2$, has the obvious definition; it is the vector space direct sum, with [ ] defined "componentwise": $[(X_1, Y_1), (X_2, Y_2)] = ([X_1X_2], [Y_1Y_2])$. The two summands $\mathfrak{g}_1$ and $\mathfrak{g}_2$ (i.e., the $(X, 0)$ and $(0, Y)$) are ideals in the direct sum that have intersection 0 and "nullify" each other ($[\mathfrak{g}_1, \mathfrak{g}_2] = 0$). Conversely, if $\mathfrak{a}$ and $\mathfrak{b}$ are two ideals in $\mathfrak{g}$ that span $\mathfrak{g}$ linearly (i.e., $\mathfrak{a}+\mathfrak{b} = \mathfrak{g}$) and have intersection 0, then the map $(X, Y) \to X + Y$ is an isomorphism of $\mathfrak{a} \oplus \mathfrak{b}$ with $\mathfrak{g}$ (thus $\mathfrak{g}$ is *internal* direct sum of $\mathfrak{a}$ and $\mathfrak{b}$). (This uses the fact that $[\mathfrak{ab}]$ is contained in $\mathfrak{a} \cap \mathfrak{b}$, and so is 0 in the present situation.) One calls $\mathfrak{a}$ and $\mathfrak{b}$ complementary ideals. An ideal $\mathfrak{a}$ is *direct summand* if there exists a complementary ideal, or, equivalently, if there exists a "retracting" homomorphism $\rho : \mathfrak{g} \to \mathfrak{a}$ with $\rho \circ i = id_{\mathfrak{a}}$(here $i : \mathfrak{a} \subset \mathfrak{g}$).

We make some comments on *change of base field*: A vector space V, or a Lie algebra $\mathfrak{g}$, over $\mathbb{C}$ can be regarded as one over $\mathbb{R}$ by *restriction of scalars*; this is the *real restriction* or *realification*, indicated by writing $V_{\mathbb{R}}$ or $\mathfrak{g}_{\mathbb{R}}$. In the other direction a V or $\mathfrak{g}$ over $\mathbb{R}$ can be made into (or, better, extended to) one over $\mathbb{C}$ by tensoring with $\mathbb{C}$ over $\mathbb{R}$; or, more elementary, by considering formal combinations $v + iw$ and $X + iY$ (with i the usual complex unit) and defining $(a + ib) \cdot (v + iw)$, $(a + ib) \cdot (X + iY)$, and $[X + iY, X' + iY']$ in the obvious way. This is the *complex extension* or *complexification*; we write $V_{\mathbb{C}}$ and $\mathfrak{g}_{\mathbb{C}}$. We call V a *real form* of $V_{\mathbb{C}}$. (A basis for V over $\mathbb{R}$ is also one for $V_{\mathbb{C}}$ over $\mathbb{C}$; same for $\mathfrak{g}$.)

A simple example: $\mathfrak{gl}(n, \mathbb{C})$ is the complexification $\mathfrak{gl}(n, \mathbb{R})_{\mathbb{C}}$ of $\mathfrak{gl}(, \mathbb{R})$. All this means is that a complex matrix M can be written uniquely as $A + iB$ with real matrices A, B.

For a slightly more complicated example: $\mathfrak{gl}(n, \mathbb{C})$ is also the complexification of the unitary Lie algebra $\mathfrak{u}(n)$. This comes about by writing any complex matrix M uniquely as $P + iQ$ with P, Q skew-Hermitean, putting $P = 1/2(M - M^*)$ and $Q = 1/2i(M + M^*)$. (This is the familiar decomposition into Hermitean plus i·Hermitean, because of "skew-Hermitean = i·Hermitean".)

Something noteworthy occurs when one complexifies a real Lie algebra that happens to be the realification of a complex Lie algebra:

Let $\mathfrak{g}$ be a Lie algebra over $\mathbb{C}$. We first define the *conjugate* $\bar{\mathfrak{g}}$ of $\mathfrak{g}$; it is a Lie algebra that is isomorphic to $\mathfrak{g}$ over $\mathbb{R}$, but multiplication by i in $\mathfrak{g}$ corresponds to multiplication by $-i$ in $\bar{\mathfrak{g}}$. One could take $\bar{\mathfrak{g}} = \mathfrak{g}$ over $\mathbb{R}$; we prefer to keep them separate, and denote by $\overline{X}$ the element of $\bar{\mathfrak{g}}$ corresponding to X in $\mathfrak{g}$. The basic rule is then $\overline{(aX)} = \bar{a} \cdot \overline{X}$.

(It happens frequently that $\bar{\mathfrak{g}}$ is isomorphic to $\mathfrak{g}$, namely when $\mathfrak{g}$ admits a *conjugate-linear automorphism* i.e., an automorphism φ over $\mathbb{R}$ such that $\varphi(aX) = \bar{a} \cdot \varphi(X)$ holds for all a and X. E.g., for $\mathfrak{sl}(n, \mathbb{C})$ such a map is simply complex conjugation of the matrix.)

In the same vein one defines the conjugate of a (complex) vector space V, denoted by $\overline{V}$. It is $\mathbb{R}$-isomorphic to V (with v in V corresponding to $\bar{v}$ in $\bar{V}$), and one has $\overline{(i \cdot v)} = -i \cdot \bar{v}$. (For $\mathbb{C}^n$ one can take "another copy" of $\mathbb{C}^n$ as the conjugate space, with $\bar{v}$ being "the conjugate" of v, i.e., obtained by taking the complex-conjugates of the components.) And—naturally—if φ is a representation of $\mathfrak{g}$ on V (all over $\mathbb{C}$), one has the conjugate representation

$\bar{\varphi}$ of $\bar{\mathfrak{g}}$ on $\overline{V}$, with $\bar{\varphi}(\overline{X})(\bar{v}) = \overline{\varphi(X)(v)}$. Finally, conjugation is clearly of order two; $\overline{\overline{V}} = V, \bar{\bar{\mathfrak{g}}} = \mathfrak{g}$, and $\bar{\bar{\varphi}} = \varphi$.

We come to the fact promised above.

PROPOSITION A. *$\mathfrak{g}_{\mathbb{R}\mathbb{C}}$ is isomorphic to the direct sum $\mathfrak{g} \oplus \bar{\mathfrak{g}}$. The isomorphism sends X in $\mathfrak{g}$ to the pair $(X, \bar{X})$.*

Proof: There are two ways to multiply elements of $\mathfrak{g}_{\mathbb{R}\mathbb{C}} = \mathfrak{g} \otimes_{\mathbb{R}} \mathbb{C}$ by the complex unit i, "on the left" and "on the right"; they are not the same since the tensor product is over $\mathbb{R}$ (The one on the right defines the structure of $\mathfrak{g}_{\mathbb{R}\mathbb{C}}$ as complex vector space.) In terms of formal combinations $X + iY$—which, to avoid confusion with the product of i and Y in $\mathfrak{g}$, we write as pairs $\{X, Y\}$—this amounts to $i \cdot \{X, Y\} = \{iX, iY\}$ (where iX is the product of i and X in $\mathfrak{g}$) and $\{X, Y\} \cdot i = \{-Y, X\}$. We consider the two subspaces U_1, consisting of all elements of the form $\{X, -iX\}$, and U_2, all $\{X, iX\}$. They are indeed complex subspaces; e.g., $\{X, -iX\} \cdot i$ equals $\{iX, X\}$, which can be written $\{iX, -i \cdot iX\}$, and is thus in U_1. They span $\mathfrak{g}_{\mathbb{R}\mathbb{C}}$ as direct sum; namely one can write $\{X, Y\}$ uniquely as $1/2\{X + iY, -iX + Y\} + 1/2\{X - iY, iX + Y\}$. One verifies that U_1 and U_2 are sub Lie algebras; furthermore the brackets between them are 0, so that they are ideals and produce a direct sum of Lie algebras. The maps $X \to 1/2\{X, -iX\}$, resp $X \to 1/2\{X, iX\}$, show that the first summand is isomorphic to $\mathfrak{g}$ and the second to $\bar{\mathfrak{g}}$: one checks that the maps preserve brackets; moreover under the first map we have $iX \to 1/2\{iX, X\}$, which equals $1/2\{X, -iX\} \cdot i$, so that the map is complex-linear, and similarly the second map turns out conjugate-linear.

Finally, for the second sentence of Proposition A we note that any X in $\mathfrak{g}$ appears as the pair $\{X, 0\}$ in $\mathfrak{g}_{\mathbb{R}\mathbb{C}}$, which can be written as $1/2\{X, -iX\} + 1/2\{X, iX\}$. $\surd$

1.5 Representations; the Killing form

We collect here some general definitions and facts on representations, and introduce the important *adjoint* representation. As noted before, a representation φ of a Lie algebra $\mathfrak{g}$ on a vector space V assigns to each X in $\mathfrak{g}$ an operator $\varphi(X)$ on V, with preservation of linearity and bracket. For $V = \mathbb{F}^n$ the $\varphi(X)$ are matrices, and we get the notion of *matrix representation.*

A representation φ is *faithful* if ker $\varphi = 0$, i.e., if the only X with $\varphi(X) = 0$ is 0 itself. If φ has kernel $\mathfrak{q}$, it induces a faithful representation of $\mathfrak{g}/\mathfrak{q}$ in the standard way. The *trivial representation* is the representation on a one-dimensional space, with all representing operators 0; as a matrix representation it assigns to each element of $\mathfrak{g}$ the matrix [0].

Let φ_1, φ_2 be two representations of $\mathfrak{g}$ on the respective vector spaces V_1, V_2. A linear map $T : V_1 \to V_2$ is *equivariant* (wr to φ_1, φ_2)), or *intertwines* φ_1 and φ_2, if it satisfies the relation $T \circ \varphi_1(X) = \varphi_2(X) \circ T$ for all X in $\mathfrak{g}$. If T is an isomorphism, then φ_1 and φ_2 are *equivalent*, and we have $\varphi_2(X) = T \circ \varphi_1(X) \circ T^{-1}$ for all X in $\mathfrak{g}$. Usually one is interested in representations only up to equivalence.

Let $\mathfrak{g}$ act on V via φ. An *invariant* or *stable* subspace is a subspace, say W, of V with $\varphi(X)(W) \subset W$ for all X in $\mathfrak{g}$. There is then an obvious induced representation of $\mathfrak{g}$ in W. Furthermore, there is an induced representation on the quotient space V/W (just as for individual operators—see Appendix), and the canonical quotient map $V \to V/W$ is equivariant.

φ and V are *irreducible* or *simple* if there is no non-trivial (i.e., different from 0 and V) invariant subspace. φ and V are *completely reducible* or *semisimple*, if every invariant subspace of V admits a complementary invariant subspace V or, equivalently, if V is direct sum of irreducible subspaces (in matrix language this means that irreducible representations are "strung along the diagonal", with 0 everywhere else).

Following the physicists's custom we will often write rep and irrep for representation and irreducible representation.

If φ is reducible (i.e., not simple), let $V_0 = 0$, $V_1 =$ a minimal invariant subspace $\neq 0$, $V_2 =$ a minimal invariant subspace containing V_1 properly, etc. After a finite number of steps one arrives at V (since $\dim V$ is finite). On each quotient V_i/V_{i-1} there is an induced simple representation; the Jordan-Hoelder theorem says that the collection of these representations is well defined up to equivalences. If φ is semisimple, then of course each V_{i-1} has a complementary invariant subspace in V_i (and conversely).

Let φ_1, φ_2 be two representations, on V_1, V_2. Their *direct sum* $\varphi_1 \oplus \varphi_2$, on $V_1 \oplus V_2$, is defined in the obvious way: $\varphi_1 \oplus \varphi_2(X)(v_1, v_2) = (\varphi_1(X)(v_1), \varphi_2(X)(v_2))$. There is also the *tensor product* $\varphi_1 \otimes \varphi_2$, on the tensor product $V_1 \otimes V_2$, defined by $\varphi_1 \otimes \varphi_2(X)(v_1 \otimes v_2) = \varphi_1(X)(v_1) \otimes v_2 + v_1 \otimes \varphi_2(X)(v_2)$. (This is the infinitesimal version of the tensor product of operators: let T_1, T_2 be operators on V_1, V_2; then, taking the derivative of $\exp(sT_1) \otimes \exp(sT_2)$ at $s = 0$, one gets $T_1 \otimes \mathrm{id} + \mathrm{id} \otimes T_2$. Note that $\varphi_1 \otimes \varphi_2(X)$ is not the tensor product of the two operators $\varphi_1(X)$ and $\varphi_2(X)$; it might be better to call it the *infinitesimal tensor product* or *tensor sum* and use some other symbol, e.g., $\varphi_1 \# \varphi_2(X)$; however, we stick with the conventional notation.) All of this extends to higher tensor powers, and also to symmetric and exterior powers of a representation (and to tensors of any kind of symmetry).

Finally, to a representation φ on V is associated the *contragredient* (strictly speaking the *infinitesimal contragredient*) or *dual* representation $\varphi^{\triangle}$ on the dual vector space $V^{\top}$, given by $\varphi^{\triangle}(X) = -\varphi(X)^{\top}$. This is a representation. The minus sign is essential; it corresponds to the fact that for the contragredient of a representation of a group one has to take the inverse of the transpose, since inverse and transpose separately yield anti-representations. And the derivative at $s = 0$ of $\exp(sT^\top)^{-1}$ is $-T^\top$.

The notions of realification and complexification of vector spaces and Lie algebras (see §1.5) extend in the obvious way to representations: From $\varphi : \mathfrak{g} \to \mathfrak{gl}(V)$ over $\mathbb{R}$ (resp. $\mathbb{C}$) we get $\varphi_\mathbb{C} : \mathfrak{g}_\mathbb{C} \to \mathfrak{gl}(V_\mathbb{C})$ (resp. $\varphi_\mathbb{R} : \mathfrak{g}_\mathbb{R} \to \mathfrak{gl}(V_\mathbb{R})$). To realify a complex representation amounts to treating a complex matrix $A + iB$ as the real matrix $\begin{bmatrix} A & -B \\ B & A \end{bmatrix}$ of twice the size. To complexify a (real) representation of a real $\mathfrak{g}$ on a real vectorspace amounts to considering real matrices as complex, via $\mathbb{R} \subset \mathbb{C}$.

The important case is that of a representation φ of a real $\mathfrak{g}$ on a complex vector space V. Here we extend φ to a representation of $\mathfrak{g}_\mathbb{C}$ on V by putting $\varphi(X + iY) = \varphi(X) + i\varphi(Y)$. This process sets up a bijection between the representations of $\mathfrak{g}$ on complex vector spaces (or by complex matrices) and the (complex!) representations of $\mathfrak{g}_\mathbb{C}$. (Both kinds of representations are determined by their values on a basis of $\mathfrak{g}$. Those of $\mathfrak{g}_\mathbb{C}$ are easier to handle because of the usual advantages of complex numbers.)

A very important representation of $\mathfrak{g}$ is the *adjoint* representation, denoted by "ad". It is just the (left) regular representation of $\mathfrak{g}$: The vector space, on which it operates, is $\mathfrak{g}$ itself; the operator $\operatorname{ad} X$, assigned to X, is given by $\operatorname{ad} X(Y) = [XY]$ for all Y in $\mathfrak{g}$ ("$\operatorname{ad} X = [X-]$"). The representation condition $\operatorname{ad}[XY] = \operatorname{ad} X \circ \operatorname{ad} Y - \operatorname{ad} Y \circ \operatorname{ad} X$ for any X, Y in $\mathfrak{g}$ turns out to be just the Jacobi condition (plus skew-symmetry). The kernel of ad is the center of $\mathfrak{g}$, as one sees immediately. Ideals of $\mathfrak{g}$ are the same as ad-invariant subspaces.

Let X be an element of $\mathfrak{g}$, and let $\mathfrak{h}$ be a sub Lie algebra (or even just a subspace), invariant under $\operatorname{ad} X$. The operator induced on $\mathfrak{h}$ by $\operatorname{ad} X$ is occasionally written $\operatorname{ad}_\mathfrak{h} X$; similarly one writes $\operatorname{ad}_{\mathfrak{g}/\mathfrak{h}} X$ for the induced operator on $\mathfrak{g}/\mathfrak{h}$. These are called the $\mathfrak{h}-$ and $\mathfrak{g}/\mathfrak{h}-$ parts of $\operatorname{ad} X$.

Remark: $\operatorname{ad} X$ is the infinitesimal version of conjugation by $\exp(sX)$, see comment (3) at the end of §1.3.

We write $\operatorname{ad} \mathfrak{g}$ for the *adjoint Lie algebra,* the image of $\mathfrak{g}$ under ad in $\mathfrak{gl}(\mathfrak{g})$.

From the adjoint representation we derive the *Killing form* κ (named after W. Killing; in the literature often denoted by B) of $\mathfrak{g}$, a symmetric bilinear form on $\mathfrak{g}$ given by

$$\kappa(X, Y) = \operatorname{tr}(\operatorname{ad} X \circ \operatorname{ad} Y) ,$$

the trace of the composition of $\operatorname{ad} X$ and $\operatorname{ad} Y$; we also write $\langle X, Y\rangle$ for this and think of $\langle \cdot, \cdot \rangle$ as a—possibly degenerate—inner product on $\mathfrak{g}$, attached to the Lie algebra structure on $\mathfrak{g}$ (in the important case of semisimple Lie algebras—see §1.7—it is non-degenerate). (The symmetry comes from the relation $\operatorname{tr}(ST) = \operatorname{tr}(TS)$ for any two operators.)

Similarly any representation φ gives rise to the symmetric bilinear *trace form* t_φ, defined by

$$t_\varphi(X, Y) = \operatorname{tr}(\varphi(X) \circ \varphi(Y)) .$$

The Killing form is *invariant* under all automorphisms of $\mathfrak{g}$: Let α be an automorphism; then we have

$$\langle \alpha(X), \alpha(Y) \rangle = \langle X, Y \rangle$$

for all X, Y in $\mathfrak{g}$. This again follows from the symmetry property of tr, and the relation $\operatorname{ad} \alpha(X) = \alpha \circ \operatorname{ad} X \circ \alpha^{-1}$ (note $\operatorname{ad} \alpha(X)(Y) = [\alpha(X)Y] = \alpha([X, \alpha^{-1}(Y)])$).

The Killing form of an ideal $\mathfrak{q}$ of $\mathfrak{g}$ is the restriction of the Killing form of $\mathfrak{g}$ to $\mathfrak{q}$ as one verifies easily. This does not hold for sub Lie algebras in general.

Example 1: $\mathfrak{sl}(2, \mathbb{C})$. We write the elements as $X = aX_+ + bH + cX_-$ (see §1.1; but we write the basis in this order, to conform with §1.11). From the brackets between the basis vectors one finds the matrix expressions

$$\operatorname{ad} H = \begin{bmatrix} 2 & 0 & 0 \\ 0 & 0 & 0 \\ 0 & 0 & -2 \end{bmatrix}, \operatorname{ad} X_+ = \begin{bmatrix} 0 & -2 & 0 \\ 0 & 0 & 1 \\ 0 & 0 & 0 \end{bmatrix}, \operatorname{ad} X_- = \begin{bmatrix} 0 & 0 & 0 \\ -1 & 0 & 0 \\ 0 & 2 & 0 \end{bmatrix}$$

and then the values $\operatorname{tr}(\operatorname{ad} H \circ \operatorname{ad} H)$ etc. of the coefficients of the Killing form, with the result

$$\kappa(X, X) = 8(b^2 + ac) \quad (= 4 \operatorname{tr} X^2) .$$

The bilinear form $\kappa(X, Y)$ is then obtained by polarization.

If we restrict to $\mathfrak{su}(2)$, by putting $b = i\alpha$ and $a = \beta + i\gamma, c = -\beta + i\gamma$, the Killing form turns into the negative definite expression $-4(\alpha^2 + \beta^2 + \gamma^2)$. For the general context, into which this fits, see §2.10.

Example 2: We consider $\mathfrak{o}(3)$ (Example 4 in §1.1), and its natural action on $\mathbb{R}^3$ (we could also use $\mathfrak{o}(3, \mathbb{C})$ and $\mathbb{C}^3$). We write the general element X as $aR_x + bR_y + cR_z$, with $a, b, c \in \mathbb{R}^3$, thus setting up an isomorphism, as vector spaces, of $\mathfrak{o}(3)$ with $\mathbb{R}^3$. Working out the adjoint representation, one finds the equations

$$\operatorname{ad} R_x = R_x, \operatorname{ad} R_y = R_y, \operatorname{ad} R_z = R_z$$

for the matrices. (In other words, the adjoint representation is equivalent to the original representation.) Computing the traces of $R_x \cdot R_x$ etc. one finds the Killing form as

$$\kappa(X, X) = -2(a^2 + b^2 + c^2).$$

Surprisingly (?) the quadratic form that defined the orthogonal Lie algebra in the first place, appears here also as the Killing form (up to a factor).

Example 3: The general linear Lie algebra $\mathfrak{gl}(n, \mathbb{F})$. Given an element A of it, the map $(\mathrm{ad}\, A)^2$ (acting on the space of all $n \times n$ matrices) sends any M to $A^2 \cdot M - 2A \cdot M \cdot A + M \cdot A^2$. One reads off from this that the Killing form, the trace of the map, is

$$\kappa(A, A) = 2n\mathrm{tr}\,(A^2) \quad - \quad 2(\mathrm{tr}\, A)^2.$$

For the special linear Lie algebra, which is an ideal in the general one, the Killing form is obtained by restriction. Thus one gets here simply $2n\mathrm{tr}\,(A^2)$.

A *derivation* of a Lie algebra $\mathfrak{g}$ is an operator $D : \mathfrak{g} \to \mathfrak{g}$ that satisfies $D[XY] = [DX, Y] + [X, DY]$ for all X, Y in $\mathfrak{g}$.
This is the infinitesimal version of automorphism: If $\alpha(s)$ is a differentiable family of automorphisms with $\alpha(0) = id$, one finds on differentiating (using Leibnitz's rule) the relation $\alpha(s)([XY]) = [\alpha(s)(X)\alpha(s)(Y)]$ that $\alpha'(0)$, the derivative at 0, is a derivation. In other words, the first order term in the expansion $\alpha(s) = id + sD + \cdots$ is a derivation. Conversely, if D is a derivation, then all $\exp(sD)$ are automorphisms, as one sees again by differentiating.

An important special case: Each $\mathrm{ad}\, X$ is a derivation of $\mathfrak{g}$; this is just the Jacobi identity; the $\mathrm{ad}\, X$'s are the *inner derivations* of $\mathfrak{g}$, analogs of the inner automorphisms of a group.

The Killing form is (infinitesimally) invariant under any derivation D of $\mathfrak{g}$, i.e., we have $\kappa(DX, Y) + \kappa(X, DY) = 0$ for all X, Y. (This is the infinitesimal version of invariance of κ under automorphisms—consider the derivative, at $s = 0$, of $\langle\alpha(s)(X), \alpha(s)(Y)\rangle = \langle X, Y\rangle$.)
The proof uses the easily verified relation $\mathrm{ad}\, DX = D \circ \mathrm{ad}\, X - \mathrm{ad}\, X \circ D$, and symmetry of tr.
Specialized to an inner derivation, this becomes the important relation

$$(*) \qquad \kappa([XY], Z) + \kappa(Y, [XZ]) = 0$$

for all X, Y, Z. I.e., $\mathrm{ad}\, X$ is skew-symmetric wr to κ.

Similarly any trace form t_φ, associated to a representation φ, is ad-invariant: $t_\varphi([XY], Z) + t_\varphi(Y, [XZ]) = 0$.

1.6 Solvable and nilpotent

The *derived* sub Lie algebra $\mathfrak{g}'$ of the Lie algebra $\mathfrak{g}$ is the ideal $[\mathfrak{g}\mathfrak{g}]$, spanned by all $[XY]$; it corresponds to the commutator subgroup of a group. The quotient $\mathfrak{g}/\mathfrak{g}'$ is Abelian, and $\mathfrak{g}'$ is the unique minimal ideal of $\mathfrak{g}$ with Abelian quotient; this is immediate from the fact that the image of $[XY]$ in $\mathfrak{g}/\mathfrak{q}$ is 0 exactly if $[XY]$ is in $\mathfrak{q}$. Clearly $\mathfrak{g}'$ is a *characteristic* ideal of $\mathfrak{g}$, that is, it is mapped into itself under every automorphism of $\mathfrak{g}$ (in fact even under any endomorphism and any derivation).

We form the *derived series*: $\mathfrak{g}, \mathfrak{g}', \mathfrak{g}'' = (\mathfrak{g}')', \ldots, \mathfrak{g}^{(r)}, \ldots$ (e.g., $\mathfrak{g}''$ is spanned by all $[[XY][UV]]$). All these $\mathfrak{g}^{(r)}$ are ideals in $\mathfrak{g}$ (in fact characteristic ones); clearly $\mathfrak{g}^{(r)} \supset \mathfrak{g}^{(r+1)}$. One calls $\mathfrak{g}$ *solvable,* if the derived series goes down to 0, i.e., if $\mathfrak{g}^{(r)}$ is 0 for large r. If $\mathfrak{g}$ is solvable, then the last non-zero ideal in the derived series is Abelian. Note: $\mathfrak{o}(3)' = \mathfrak{o}(3)$, thus $\mathfrak{o}(3)$ is not solvable; $\mathfrak{aff}(1)'' = 0$, so $\mathfrak{aff}(1)$ is solvable. The prime example for solvability is formed by the Lie algebra of upper-triangular matrices ($a_{ij} = 0$ for $i > j$).

The *lower central series*, $\mathfrak{g}, \mathfrak{g}^1, \mathfrak{g}^2, \ldots, \mathfrak{g}^r, \ldots$ is defined inductively by $\mathfrak{g}^1 = \mathfrak{g}', \mathfrak{g}^{r+1} = [\mathfrak{g}, \mathfrak{g}^r]$; thus $\mathfrak{g}^r$ is spanned by *iterated* or *long* brackets $[X_1[X_2[\ldots X_{r+1}]\ldots]$ (which we abbreviate to $[X_1X_2\ldots X_{r+1}]$). Again the $\mathfrak{g}^r$ are characteristic ideals, and the relation $g^{r+1} \subset \mathfrak{g}^r$ holds. One calls $\mathfrak{g}$ *nilpotent,* if the lower central series goes down to 0, i.e., if $\mathfrak{g}^r$ is 0 for large r. The standard example for nilpotence are the upper supra-triangular matrices, those with $a_{ij} = 0$ for $i \geq j$. (This is the derived Lie algebra of the upper-triangular one.)

One sees easily that the derived and lower central series of an ideal of $\mathfrak{g}$ consists of ideals of $\mathfrak{g}$.

Nilpotency implies solvability, because of the relation $\mathfrak{g}^{(r)} \subset \mathfrak{g}^r$ (easily proved by induction); the converse is not true—consider $\mathfrak{aff}(1)$. It is also fairly clear that a sub Lie algebra of a solvable (resp nilpotent) Lie algebra is itself solvable (resp nilpotent), and similar for quotients. For solvability there is a "converse":

LEMMA A. *Let $0 \to \mathfrak{q} \to \mathfrak{g} \to \mathfrak{p} \to 0$ be an exact sequence of Lie algebras. Then $\mathfrak{g}$ is solvable iff both $\mathfrak{q}$ and $\mathfrak{p}$ are so.*

In one direction we have seen this already. For the other, note that $\mathfrak{g}^{(r)}$ maps into $\mathfrak{p}^{(r)}$; the latter is 0 for large r, and so $\mathfrak{g}^{(r)}$ is contained in the image of $\mathfrak{q}$. Then $\mathfrak{g}^{(r+s)}$ is in the image of $\mathfrak{q}^{(s)}$; and the latter is 0 for large s. $\surd$

We show next that $\mathfrak{g}$ contains a unique maximal solvable ideal (i.e., there is such an ideal that contains all solvable ideals), the *radical* $\mathfrak{r}$ of $\mathfrak{g}$; similarly there is a unique maximal nilpotent ideal, occasionally called the *nilradical* $\mathfrak{n}$. This is an immediate consequence of the following

LEMMA B. *If $\mathfrak{a}$ and $\mathfrak{b}$ are solvable (resp. nilpotent) ideals of $\mathfrak{g}$, then so is the ideal $\mathfrak{a}+\mathfrak{b}$.*

Proof: For the solvable case we have the exact sequence $0 \to \mathfrak{a} \to \mathfrak{a}+\mathfrak{b} \to (\mathfrak{a}+\mathfrak{b})/\mathfrak{a} \to 0$; the third term is isomorphic to $\mathfrak{b}/\mathfrak{a}\cap\mathfrak{b}$ and so solvable, and we can apply Lemma A. For the nilpotent case one verifies that any long bracket with $s+1$ of its terms in $\mathfrak{a}$ lies in $\mathfrak{a}^s$; for example $[a_1[a_2b]]$ is in $\mathfrak{a}^1$, because $[a_2b]$ is in $\mathfrak{a}$. Therefore all sufficiently long brackets of $\mathfrak{a}+\mathfrak{b}$ are 0, since they belong either to $\mathfrak{a}^s$ with large s or to $\mathfrak{b}^t$ with large t. $\surd$

The nilradical is of course contained in the radical.

We come to a fundamental definition, singling out a very important class of Lie algebras: A Lie algebra $\mathfrak{g}$ is called *semisimple*, if its radical is 0 and its dimension is positive. (Since the last term of the derived series is an Abelian ideal, vanishing of the radical amounts to the same as: if there is no non-zero Abelian ideal.)

From Lemma A it follows that the quotient $\mathfrak{g}/\mathfrak{r}$ of $\mathfrak{g}$ by its radical $\mathfrak{r}$ is semisimple; thus in a sense (i.e., up to *extensions*), semisimple and solvable Lie algebras yield all Lie algebras (see the Levi-Malcev theorem below). The quotient of $\mathfrak{g}$ by its nilradical $\mathfrak{n}$ may well have a non-zero nilradical; example: $\mathfrak{aff}(1)$.

The importance of semisimplicity comes from its equivalence (§1.10, Theorem A) with the non-degeneracy of the Killing form of $\mathfrak{g}$.

One more basic definition: A Lie algebra $\mathfrak{g}$ is *simple*, if it has no non-trivial ideals (different from 0 or $\mathfrak{g}$) and is not of dimension 0 or 1.

[The dimension restriction only excludes the rather trivial Abelian Lie algebra of dimension one; it is actually equivalent to requiring $\mathfrak{g}$ not Abelian, or to requiring $\mathfrak{g}$ semisimple: If $\mathfrak{g}$ has dimension greater than 1, it is not Abelian (otherwise it would have non-trivial ideals). If it is not Abelian, it is not solvable (the absence of non-trivial ideals would make it Abelian); thus the radical is a proper ideal (i.e., $\neq \mathfrak{g}$) and so equal to 0, making $\mathfrak{g}$ semisimple. And if $\mathfrak{g}$ is semisimple, it must be of dimension more than 1 anyway.]

We shall soon prove the important fact that every semisimple Lie algebra is direct sum of simple ones, and we shall later (in Ch.2) find all simple Lie algebras (over $\mathbb{C}$). As for solvable Lie algebras, although a good many general facts are known, there is no complete list of all possibilities. For the "general" Lie algebra, we have the exact sequence $0 \to \mathfrak{r} \to \mathfrak{g} \to \mathfrak{g}/\mathfrak{r} \to 0$, with $\mathfrak{r}$ solvable and $\mathfrak{g}/\mathfrak{r}$ semisimple. Furthermore there is the Levi-Malcev theorem (which we shall not prove, although it is not difficult) that this sequence splits, i.e., that $\mathfrak{g}$ contains a sub Lie algebra complementary to $\mathfrak{r}$

(and so isomorpic to $\mathfrak{g}/\mathfrak{r}$). Thus every Lie algebra is put together from a solvable and a semisimple part. We describe how the two parts interact:

If one analyzes the brackets between elements of $\mathfrak{r}$ and $\mathfrak{s}$, one is led to the notion of *semidirect sum:* Let $\mathfrak{a}, \mathfrak{b}$ be two Lie algebras, and let there be given a representation φ of $\mathfrak{a}$ on (the vector space) $\mathfrak{b}$ by *derivations* of $\mathfrak{b}$ (i.e., every $\varphi(X)$ is a derivation of $\mathfrak{b}$). We make the vector space direct sum of $\mathfrak{a}$ and $\mathfrak{b}$ into a Lie algebra, denoted by $\mathfrak{a} \oplus_\varphi \mathfrak{b}$, by using the given brackets in the two summands $\mathfrak{a}$ and $\mathfrak{b}$, and by defining $[XY] = \varphi(X)(Y)$ for X in $\mathfrak{a}$ and Y in $\mathfrak{b}$. This is indeed a Lie algebra (the derivation property of the $\varphi(X)$'s is of course essential here), and there is an exact sequence $0 \to \mathfrak{b} \to \mathfrak{a} \oplus_\varphi \mathfrak{b} \to \mathfrak{a} \to 0$, which is in fact split, via the obvious embedding of $\mathfrak{a}$ as the first summand of $\mathfrak{a} \oplus \mathfrak{b}$. (For $\varphi = 0$ this gives the ordinary direct sum.) In these terms then, the general $\mathfrak{g}$ is semidirect sum of a semisimple Lie algebra $\mathfrak{s}$ and a solvable Lie algebra $\mathfrak{r}$, under some representation of $\mathfrak{s}$ on $\mathfrak{r}$ by derivations.

1.7 Engel's theorem

We begin the more detailed discussion of Lie algebras with a theorem that, although it is rather special, is technically important; it is known as *Engel's theorem.* It connects nilpotence of a Lie algebra with ordinary nilpotence of operators on a vector space.

THEOREM A. *Let V be a vector space; let $\mathfrak{g}$ be a sub Lie algebra of the general linear Lie algebra $\mathfrak{gl}(V)$, consisting entirely of nilpotent operators. Then $\mathfrak{g}$ is a nilpotent Lie algebra.*

Second form of Engel's theorem:

THEOREM A′. *If $\mathfrak{g}$ is a Lie algebra such that all operators $\operatorname{ad} X$, with X in $\mathfrak{g}$, are nilpotent, then $\mathfrak{g}$ is nilpotent.*

For the proof we start with

PROPOSITION B. *Let the Lie algebra $\mathfrak{g}$ act on the non-zero vector space V by nilpotent operators; then the nullspace*

$$N = \{v \in V : Xv = 0 \text{ for all } X \text{ in } \mathfrak{g}\}$$

is not 0.

We prove this by induction on the dimension of $\mathfrak{g}$ (most theorems on nilpotent and solvable Lie algebras are proved that way). The case $\dim \mathfrak{g} = 0$ is clear. Suppose the proposition holds for all dimensions $< n$, and take $\mathfrak{g}$

of dimension n (> 0). We may assume the representation φ at hand faithful, since otherwise the effective Lie algebra $\mathfrak{g}/\ker\varphi$ has dimension $< n$. Thus we can consider $\mathfrak{g}$ as sub Lie algebra of $\mathfrak{gl}(V)$. Now $\mathfrak{g}$ operates on itself (actually on all of $\mathfrak{gl}(V)$) by ad; and all operators $\operatorname{ad} X$, for X in $\mathfrak{g}$, are nilpotent: We have $\operatorname{ad} X.Y = XY - YX, (\operatorname{ad} X)^2.Y = X^2Y - 2XYX + YX^2, \ldots$, and the factors X pile up on one side or the other. (If $X^k = 0$, then $(\operatorname{ad} X)^{2k} = 0$.) Let $\mathfrak{m}$ be a maximal sub Lie algebra of $\mathfrak{g}$ different from $\mathfrak{g}$ (sub Lie algebras $\neq \mathfrak{g}$ exist, e.g. 0; take one of maximal dimension). $\mathfrak{m}$ operates on $\mathfrak{g}$ by restriction of ad.

This operation leaves $\mathfrak{m}$ invariant, since $\mathfrak{m}$ is a sub Lie algebra, and so there is the induced representation in $\mathfrak{g}/\mathfrak{m}$. This representation is still by nilpotent operators, and thus the null space is non-zero, by induction hypothesis. A non-zero element in this subspace is represented by an element X_0 not in $\mathfrak{m}$. The fact that X_0 is nullified modulo $\mathfrak{m}$ by $\mathfrak{m}$ translates into $[\mathfrak{m}X_0] \subset \mathfrak{m}$. Thus $((\mathfrak{m}, X_0))$ is a sub Lie algebra of $\mathfrak{g}$, which by maximality of $\mathfrak{m}$ must be equal to $\mathfrak{g}$.

By induction hypothesis the nullspace U of $\mathfrak{m}$ in the original V is non-zero; and the operator relation $YX_0 = X_0Y + [YX_0]$ shows that X_0 maps U into itself (if u is nullified by all Y in $\mathfrak{m}$, so is X_0u: apply both sides of the relation to u and note that $[YX_0]$ is in $\mathfrak{m}$). The operator X_0 is still nilpotent on U and so has a non-zero nullvector v; and then v is a non-zero nullvector for all of $\mathfrak{g}$. $\surd$

We now prove Theorem A. We apply Proposition B to the contragredient action of $\mathfrak{g}$ on the dual vectorspace $V^\top$ (see §1.5); the operators are of course nilpotent. We find a non-zero linear function λ on V that is annulled by $\mathfrak{g}$. It follows that the space $((\mathfrak{g} \cdot V))$, spanned by all Xv with X in $\mathfrak{g}$ and v in V, is a proper subspace of V; namely it is contained in the kernel of λ, by $\lambda(Xv) = X^\top\lambda(v) = 0$. Since $((\mathfrak{g} \cdot V))$ is of course invariant under $\mathfrak{g}$, we can iterate the argument, and find that, with $t = \dim V$, all operators of the form $X_1 \cdot X_2 \cdot \ldots \cdot X_t$ vanish, since each X_i decreases the dimension by at least 1. This implies Engel's theorem, once we observe that any long bracket $[X_1X_2 \ldots X_k]$ expands, by $[XY] = XY - YX$, into a sum of products of k X's. The second form of Engel's theorem, Theorem A′, follows readily: taking $\mathfrak{g}$ as V and letting $\mathfrak{g}$ act by ad, we just saw that $\operatorname{ad} X_1 \cdot \operatorname{ad} X_2 \cdot \ldots \cdot \operatorname{ad} X_n$ is 0 (with $n = \dim \mathfrak{g}$), and so $[X_1X_2 \ldots X_{n+1}] = 0$ for all choices of the X's. (We remark that Engel's theorem, in contrast to the following theorems, holds for fields of any characteristic.) $\surd$

1.8 Lie's theorem

There are several equivalent forms of the theorem that commonly goes by this name:

THEOREM A. *Let $\mathfrak{g}$ be a solvable Lie algebra, acting on the vector space V by a representation φ, all over $\mathbb{C}$. Then there exists a "joint eigenvector"; i.e., there is a non- zero vector v_0 in V that satisfies $Xv_0 = \lambda(X)v_0$, where $\lambda(X)$ is a complex number (depending on X), for all X in $\mathfrak{g}$.*

$\lambda(X)$ depends of course linearly on X; i.e., λ is a linear function on V.

THEOREM A$'$. *A complex irreducible representation of a complex solvable Lie algebra is of dimension ≤ 1.*

THEOREM A$''$. *Any complex representation of a complex solvable Lie algebra is equivalent to a triangular one, i.e., to one with all matrices (upper-) triangular.*

It is easily seen that the three forms are equivalent. Note that every representation of positive dimension has non-irreducible stable subspaces (those of minimal positive dimension), and so A$'$ implies A. By considering induced representations in quotients of invariant subspaces one gets A$''$.

There is also a real version; we state the analog of A$'$.

THEOREM B. *A real irreducible representation of a real solvable Lie algebra is of dimension ≤ 2, and is Abelian (all operators commute).*

This follows from the complex version by complexification. An eigenvector $v + iw$ gives rise to the real invariant subspace $((v, w))$; the Abelian property comes from the fact that one-dimensional complex representations are Abelian.

For the proof of Lie's theorem we start with a lemma (Dynkin):

LEMMA C. *Let $\mathfrak{g}$ be a Lie algebra, acting on a vector space V; let $\mathfrak{a}$ be an ideal of $\mathfrak{g}$, and let λ be a linear function on $\mathfrak{a}$. Let W be the subspace of V spanned by all the joint eigenvectors of $\mathfrak{a}$ with eigenvalue λ (i.e., the v with $Xv = \lambda(X)v$ for X in $\mathfrak{a}$). Then W is invariant (under all of $\mathfrak{g}$).*

Proof: For v in W, A in $\mathfrak{a}$, and X in $\mathfrak{g}$ we have

$$AXv = XAv + [AX]v = \lambda(A)Xv + \lambda([AX])v\ .$$

(Note that $[AX]$ is in $\mathfrak{a}$.) Thus to show that Xv is in W, it is sufficient to show $\lambda([AX]) = 0$. With fixed X and v we form the vectors $v_0 = v, v_1 = Xv, v_2 = X^2v, \ldots, v_i = X^iv, \ldots$ and the increasing sequence of spaces $U_i = ((v_0, v_1, \ldots, v_i))$ for $i \geq 0$. Let k be the smallest of the i with $U_i = U_{i+1}$ (this exists of course). We show inductively that all U_i are invariant under every A in $\mathfrak{a}$, and that the matrix of A on U_k is triangular wr to the basis $\{v_0, v_1, \ldots, v_k\}$, with all diagonal elements equal to $\lambda(A)$. For $i = 0$

we have $Av_0 = \lambda(A)v_0$ by hypothesis. For $i > 0$ we have $Av_i = AX^iv = XAX^{i-1}v + [AX]X^{i-1}v = XAv_{i-1} + [AX]v_{i-1}$. The second term is in U_{i-1} by induction hypothesis ($[AX]$ is in $\mathfrak{a}$). For the first term we have $Av_{i-1} = \lambda(A)v_{i-1} \bmod U_{i-2}$, and thus $XAv_{i-1} = \lambda(A)v_i \bmod U_{i-1}$. Altogether, $Av_i = \lambda(A)v_i \bmod U_{i-1}$, which clearly proves our claim. Taking trace on U_k we find $\operatorname{tr} A = (k+1)\cdot\lambda(A)$; in particular $\operatorname{tr}[AX] = (k+1)\cdot\lambda([AX])$. But U_k is clearly also invariant under X, and so $\operatorname{tr}[AX] = \operatorname{tr}(AX - XA) = 0$. With $k + 1 > 0$ this shows $\lambda([AX]) = 0$. (Note: the fact that the characteristic of the field is 0 is crucial here.) $\surd$

The proof of Lie's theorem proceeds now by induction on the dimension of $\mathfrak{g}$, the case $\dim \mathfrak{g} = 0$ being obvious. Consider a $\mathfrak{g}$ of $\dim = n(> 1)$, and suppose the theorem true for all dimensions $< n$. In $\mathfrak{g}$ there exists an ideal $\mathfrak{a}$ of codimension 1 (since any subspace containing $\mathfrak{g}'$ is an ideal, by $[\mathfrak{a}\mathfrak{g}] \subset [\mathfrak{g}\mathfrak{g}] \subset \mathfrak{a}$, with, incidentally, Abelian quotient $\mathfrak{g}/\mathfrak{a}$). By induction hypothesis $\mathfrak{a}$ has a joint eigenvector in V, with eigenvector a linear function λ. By Dynkin's lemma the space W, spanned by all eigenvectors of $\mathfrak{a}$ to λ, is invariant under $\mathfrak{g}$. Let X_0 be an element of $\mathfrak{g}$ not in $\mathfrak{a}$; we clearly have $\mathfrak{a} + ((X_0)) = \mathfrak{g}$. Since $X_0W \subset W$ and we are over $\mathbb{C}$, X_0 has an eigenvector v_0 in W, with eigenvalue λ_0 (note that by its construction W is not 0). And now v_0 is joint eigenvector for $\mathfrak{g}$, with eigenvalue $\lambda(A) + r\lambda_0$ for $X = A + rX_0$. $\surd$

1.9 Cartan's first criterion

This criterion is a condition for solvability in terms of the Killing form:

THEOREM A. *A Lie algebra $\mathfrak{g}$ is solvable iff its Killing form κ vanishes identically on the derived Lie algebra $\mathfrak{g}'$.*

It is easy to see that both solvability and vanishing of κ on $\mathfrak{g}'$ remain unchanged under complexification for a real $\mathfrak{g}$; thus we may take $\mathfrak{g}$ complex. We begin with a proposition that contains the main argument:

PROPOSITION B. *Let $\mathfrak{g}$ be a sub Lie algebra of $\mathfrak{gl}(V)$ for a vector space V with the property $\operatorname{tr}(XY) = 0$ for all X, Y in $\mathfrak{g}$. Then the derived Lie algebra $\mathfrak{g}'$ is nilpotent.*

Note that the combination XY, and not $[XY]$, appears here. The proof uses the Jordan form of operators. Take X in $\mathfrak{g}'$; we have $X = S + N$ with $SN = NS$, N nilpotent, and S diagonal $= \operatorname{diag}(\lambda_1, \ldots, \lambda_n)$ relative to a suitable basis of V. (We consider all operators on V as matrices wr to this basis and take the usual matrix units E_{ij}, with 1 as ij-entry and 0 everywhere else, as basis for $\mathfrak{gl}(V)$.) Put $\overline{S} = \operatorname{diag}(\overline{\lambda}_1, \ldots, \overline{\lambda}_n)$ (i.e., the complex conjugate of S); then $\overline{S}$ can be written as a polynomial in S, by Lagrange interpolation (since $\lambda_i = \lambda_j$ implies $\bar{\lambda}_i = \bar{\lambda}_j$, there is a polynomial $p(x)$ with $p(\lambda_i) = \bar{\lambda}_i$).

Now consider the representation ad of $\mathfrak{gl}(V)$, restricted to $\mathfrak{g}$. We have $\operatorname{ad} X = \operatorname{ad} S + \operatorname{ad} N$. Here $[SN] = 0$ implies $[\operatorname{ad} S \operatorname{ad} N] = 0$ (ad is a representation!); $\operatorname{ad} N$ is nilpotent (as in the proof of Engel's theorem); and finally, $\operatorname{ad} S$ is diagonal, with eigenvalue $\lambda_i - \lambda_j$ on E_{ij}, and so semisimple. Thus $\operatorname{ad} S + \operatorname{ad} N$ is the Jordan decomposition of $\operatorname{ad} X$; and so $\operatorname{ad} S$ is a polynomial in $\operatorname{ad} X$. Furthermore, $\operatorname{ad} \overline{S}$ is also diagonal, with eigenvalue $\bar{\lambda}_i - \bar{\lambda}_j$ on E_{ij}; therefore again $\operatorname{ad} \overline{S}$ is a polynomial in $\operatorname{ad} S$, and then also one in $\operatorname{ad} X$. This finally implies $\operatorname{ad} \overline{S}(\mathfrak{g}) \subset \mathfrak{g}$, or: $[\overline{S}Y]$ is in $\mathfrak{g}$ for Y in $\mathfrak{g}$.

From $\overline{S} = p(S)$ we infer that $\overline{S}$ and N commute, and so the product $\overline{S}N$ is nilpotent, and in particular has trace 0. Therefore we have $\operatorname{tr} \overline{S}X = \operatorname{tr} \overline{S}S = \Sigma \lambda_i \bar{\lambda}_i$.

On the other hand we have $X = \Sigma[A_r B_r]$ with A_r, B_r in $\mathfrak{g}$, since X is in $\mathfrak{g}'$; for each term we have $\operatorname{tr} \overline{S}[AB] = \operatorname{tr}(\overline{S}AB - \overline{S}BA) = \operatorname{tr} \overline{S}AB - \operatorname{tr} A\overline{S}B = \operatorname{tr}[\overline{S}A]B$, and, since $[\overline{S}A]$ is in $\mathfrak{g}$ as shown above, this vanishes by hypothesis on $\mathfrak{g}$. Thus we have $\Sigma \lambda_i \bar{\lambda}_i = 0$, which forces all λ_i to vanish, so that finally S is 0. We have shown now that all X in $\mathfrak{g}'$ are nilpotent; Engel's theorem tells us that then $\mathfrak{g}'$ is nilpotent. $\checkmark$

Now to Cartan's first criterion: Consider the representation ad of $\mathfrak{g}$ on $\mathfrak{g}$. The image is a sub Lie algebra $\mathfrak{q}$ of $\mathfrak{gl}(\mathfrak{g})$, and there is the exact sequence $0 \to z \to \mathfrak{g} \to \mathfrak{q} \to 0$, with $\mathfrak{z}$ the center of $\mathfrak{g}$ (which is solvable, even Abelian). The vanishing of the Killing form of $\mathfrak{g}$ on $\mathfrak{g}'$ translates into $\operatorname{tr} AB = 0$ for all A, B in $\mathfrak{q}'$. Proposition B gives nilpotence of $\mathfrak{q}''$, which makes $\mathfrak{q}'$ and $\mathfrak{q}$ solvable. From Lemma A, §1.6, on short exact sequences of solvable Lie algebras we find that $\mathfrak{g}$ is solvable. $\checkmark$

For the converse part of Theorem A we apply Lie's theorem to the adjoint representation. The matrices for the $\operatorname{ad} X$ are then triangular. For X in $\mathfrak{g}'$ all diagonal elements of $\operatorname{ad} X$ are then 0 (clear for any $\operatorname{ad} A \cdot \operatorname{ad} B - \operatorname{ad} B \cdot \operatorname{ad} A$); the same is then true for $\operatorname{ad} X \cdot \operatorname{ad} Y$ with X, Y in $\mathfrak{g}'$, and thus the Killing form (the trace) vanishes, in fact "quite strongly", on $\mathfrak{g}'$. $\checkmark$

1.10 Cartan's second criterion

This describes the basic connection between semisimplicity and the Killing form:

THEOREM A. *A Lie algebra $\mathfrak{g}$ is semisimple iff its dimension is positive and its Killing form is non-degenerate.*

(κ non-degenerate means: If for some X_0 in $\mathfrak{g}$ the value $\kappa(X_0, Y)$ is 0 for all Y in $\mathfrak{g}$, then X_0 is 0.)

Just as for the first criterion we may assume that $\mathfrak{g}$ is complex, since both semisimplicity and non-degeneracy of κ are unchanged by complexification (the radical of the complexification is the complexification of the radical; one can describe non-degeneracy of κ as: If $\{X_1, \ldots, X_n\}$ is a basis for $\mathfrak{g}$, then the determinant of the matrix $[\kappa(X_i, X_j)]$ is not 0).

Proof of Theorem A: (1) Suppose $\mathfrak{g}$ not semisimple. It has then a non-zero Abelian ideal $\mathfrak{a}$. Take A in $\mathfrak{a}$, not 0, and take any X in $\mathfrak{g}$. Then $\operatorname{ad} A \cdot \operatorname{ad} X \cdot \operatorname{ad} A$ maps $\mathfrak{g}$ into 0 (namely $\mathfrak{g} \to \mathfrak{a} \to \mathfrak{a} \to 0$), and $\operatorname{ad} A \cdot \operatorname{ad} X$ is nilpotent (of order 2). So $\kappa(A, X)$, the trace of $\operatorname{ad} A \cdot \operatorname{ad} X$, is 0, and κ is degenerate.

(2) Suppose κ degenerate. Put $\mathfrak{g}^\perp = \{X : \kappa(X, Y) = 0 \text{ for all } Y \text{ in } \mathfrak{g}\}$; this is the *degeneracy subspace* or *radical* of κ; it is not 0, by assumption. It is also an ideal, as follows from the (infinitesimal) invariance of κ (we have $\kappa(X, [YZ]) = \kappa([XY], Z), by (*) in §1.5)$, and so $[XY]$ is in $\mathfrak{g}^\perp$, if X is. Obviously the restriction of κ to $\mathfrak{g}^\perp$ is identically 0. Since the restriction of the Killing form to an ideal is the Killing form of the ideal, the Killing form of $\mathfrak{g}^\perp$ is 0. Cartan's first criterion then implies that $\mathfrak{g}^\perp$ is solvable, and so $\mathfrak{g}$ is not semisimple. $\surd$

There are three important corollaries.

COROLLARY B. *A Lie algebra $\mathfrak{g}$ is semisimple iff it is direct sum of simple Lie algebras.*

Let $\mathfrak{g}$ be semisimple, and let $\mathfrak{a}$ be any (non-zero) ideal. Then $\mathfrak{a}^\perp = \{X : \kappa(X, Y) = 0 \text{ for all } Y \text{ in } \mathfrak{a}\}$ is also an ideal, by the invariance of κ, as above. Non-degeneracy of κ implies $\dim \mathfrak{a} + \dim \mathfrak{a}^\perp = \dim \mathfrak{g}$. (If $\{Y_1, \ldots, Y_r\}$ is a basis of $\mathfrak{a}$, then the equations $\kappa(X, Y_1) = 0, \ldots, \kappa(X, Y_r) = 0$ are independent). Furthermore $\mathfrak{a} \cap \mathfrak{a}^\perp$ is also an ideal of $\mathfrak{g}$, with vanishing Killing form (arguing as above), therefore solvable (by Cartan's first criterion), and therefore 0 by semisimplicity of $\mathfrak{g}$. It follows that $\mathfrak{g}$ is the direct sum of $\mathfrak{a}$ and $\mathfrak{a}^\perp$ (note $[\mathfrak{a}, \mathfrak{a}^\perp]$ is 0, as sub Lie algebra of $\mathfrak{a} \cap \mathfrak{a}^\perp$). Clearly $\mathfrak{a}$ and $\mathfrak{a}^\perp$ must be semisimple (they can't have solvable ideals, or, their Killing forms must be non-degenerate). Thus we can use induction on the dimension of $\mathfrak{g}$. $\surd$

The argument in the other direction is simpler: semisimplicity is preserved under direct sum, and simple implies semisimple.

COROLLARY C. *A semisimple ideal in a Lie algebra is direct summand.*

The proof is substantially the same as that for Corollary B. The complementary ideal is found as the subspace orthogonal to the ideal wr to the Killing form. The intersection of the two is 0, since by Cartan's first criterion it is a solvable ideal in the given ideal. $\surd$

COROLLARY D. *Every derivation of a semisimple Lie algebra is inner.*

Let $\mathfrak{g}$ be the Lie algebra and D the derivation. In the vector space $\mathfrak{g} \oplus \mathbb{F}D$, spanned by $\mathfrak{g}$ and the abstract "vector" D, we define a [ ]-operation by $[DD] = 0, [DX] = -[XD] = DX$ (i.e., equal to the image of X under D), and the given bracket within $\mathfrak{g}$. One checks that this is a Lie algebra, and that it has $\mathfrak{g}$ as an ideal. By Corollary C there is a complementary ideal, which is of dimension 1 and is clearly spanned by an element of the form $-X_0 + D$, with some X_0 in $\mathfrak{g}$. Complementarity implies $[-X_0 + D, X] = 0$, i.e., $DX = \operatorname{ad} X_0.X$ for all X in $\mathfrak{g}$; in short, $D = \operatorname{ad} X_0$. $\surd$

1.11 Representations of A_1

From §1.1 we recall that $A_1, = \mathfrak{sl}(2, \mathbb{C})$, is the (complex) Lie algebra with basis $\{H, X_+, X_-\}$ and relations

$$[HX_+] = 2X_+, [HX_-] = -2X_-, [X_+X_-] = H.$$

(Incidentally, this is also $\mathfrak{su}(2)_{\mathbb{C}}$, the complexified $\mathfrak{su}(2)$, and therefore also $\mathfrak{o}(3)_{\mathbb{C}}$. Indeed, H, X_+, X_- are equal to, respectively, $-2iS_z, -iS_x - S_y, -iS_x + S_y$, with the S's of §1.1, Example 9.)

Our purpose in this and the following section is to describe all representations of A_1. We do this here, in order to have something concrete to look at and also because the facts are of general interest (e.g., in physics, in particular in elementary quantum theory); furthermore, the results foreshadow the general case; and, finally, we will use the results in studying the structure and representations of semisimple Lie algebras.

Let then an action of A_1 on a (complex) vectorspace V be given. The basis of all the following arguments is the following simple fact:

LEMMA A. *Let v be an eigenvector of (the operator assigned to) H, with eigenvalue λ. Then X_+v and X_-v, if different from 0, are also eigenvectors of H, with eigenvalues $\lambda + 2$ and $\lambda - 2$.*

Proof. We are given $Hv = \lambda v$. In the language of physics, we "use the commutation relations", i.e., we note that $[HX_+]$ acts as $H \circ X_+ - X_+ \circ H$. Thus we have $HX_+v = X_+Hv + [HX_+]v = \lambda X_+v + 2X_+v = (\lambda + 2)X_+v$; similarly for X_-. $\surd$

To analyze the action of A_1, we first note that eigenvectors of H exist, of course (that is the reason for using $\mathbb{C}$). Take such a one, v, and form the sequence $v, X_+v, (X_+)^2v, \ldots$ (iterating X_+). By Lemma A all these vectors are either 0 or eigenvectors of H, with no two belonging to the same eigenvalue. Since H has only a finite number of eigenvalues, we will arrive at a non-zero vector v_0 that satisfies $Hv_0 = \lambda v_0$ for some λ and $X_+v_0 = 0$. With this v_0 we define $v_1 = X_-v_0, v_2 = X_-v_1, \ldots$ (iterating X_-); we also define $v_{-1} = 0$. Let v_r be the last non-zero vector in the sequence.

By Lemma A we have $Hv_i = (\lambda - 2i)v_i$ for all $i \geq -1$. Next we prove, inductively, the relations $X_+v_i = \mu_i v_{i-1}$ with $\mu_i = i \cdot (\lambda+1-i)$, for all $i \geq 0$. The case $i = 0$ is clear, with $\mu_0 = 0$. The induction step consists in the computation $X_+v_{i+1} = X_+X_-v_i = X_-X_+v_i + [X_+X_-]v_i = \mu_i v_i + Hv_i = (\mu_i + \lambda - 2i)$, which shows $\mu_{i+1} = \mu_i + \lambda - 2i$; with the initial condition $\mu_0 = 0$ this gives the claimed value for μ_i. Now we take $i = r+1$, so that $v_r \neq 0$, but $v_{r+1} = 0$. From $0 = X_+v_{r+1} = \mu_{r+1}v_r$ we read off $\mu_{r+1} = 0$; this gives $\lambda = r$.

The vectors $v_0, v_1, \ldots, v_r$ are eigenvectors of H to different eigenvalues and so independent. The formulae for the action of X_+ and X_- show that the space $((v_0, v_1, \ldots, v_r))$ is invariant under the action of A_1. [In fact, the action is very simple: X_+ moves the v_i "down", X_- moves them "up", and the "ends" go to 0.] In particular, if V irreducible, this space is equal to V. Thus we know what irreducible representations must look like.

It is also clear that irreducible representations of this type exist. Take any natural number $r \geq 0$. Take a vector space of dimension $r+1$, with a basis $\{v_0, v_1, \ldots, v_r\}$, and define an action of A_1 by the formulae above: $Hv_i = (r-2i)v_i, X_-v_i = v_{i+1}$ (and $= 0$ for $i = r$), $X_+v_i = \mu_i v_{i-1}$ with $\mu_i = i(r+1-i)$ (and $= 0$ for $i = 0$). It should be clear that this is indeed a representation of A_1, i.e., that the relations $[X_+X_-]v = Hv$, etc., hold for all vectors v in the space.

Furthermore, this representation is irreducible: From any non-zero linear combination of the v_i one gets, by a suitable iteration of X_+, a non-zero multiple of v_0, and then, with the help of X_-, all the v_i. $\surd$

It is customary to put $r = 2s$ (with $s = 0, 1/2, 1, \ldots$), and to denote the representation just described by D_s. It is of dimension $2s+1$. We write out the matrices for H, X_+, X_- under D_s, wr to the v_i-basis. The μ_i, $= i(2s+1-i)$, strictly speaking should carry s as a second index.

$$H \to \operatorname{diag}(2s, 2s-2, \ldots, 2-2s, -2s)$$

$$X_+ \to \begin{bmatrix} 0 & \mu_1 & & & 0 \\ & 0 & \mu_2 & & \\ & & \ddots & & \\ & & & 0 & \mu_r \\ 0 & & & & 0 \end{bmatrix}, X_- \to \begin{bmatrix} 0 & & & & 0 \\ 1 & 0 & & & \\ & 1 & 0 & & \\ & & & \ddots & \\ 0 & & & 1 & 0 \end{bmatrix}$$

We emphasize: H (i.e., the matrix representing it in D_s) is diagonal; the eigenvalues are integers; they range in steps of 2 from $2s$ to $-2s$. As for X_+ and X_-, the shape of the matrix (off-diagonal) is fixed; but the entries (in contrast to those for H) change, in a simple way, if one modifies the v_i by numerical factors. The following normalization is fairly common in physics: The basic vectors are called v_m, with m running down in steps of one from s to $-s$ with $Hv_m = 2m \cdot v_m$. The two other operators are defined by $X_+ v_m = \sqrt{s(s+1) - m(m+1)} \cdot v_{m+1}, X_- v_m = \sqrt{s(s+1) - m(m-1)} \cdot v_{m-1}$. (The value $s(s+1) - m(m+1)$ corresponds to our earlier $i(2s+1-i)$.)

We have established the *classification* result:

THEOREM B. *The representations D_s, with $s = 0, 1/2, 1, 3/2, \ldots$, of dimension $2s+1$, form the complete list (up to equivalence) of irreducible representations of A_1.*

We note: D_0 is the *trivial representation*, of dimension 1 (all operators are 0). $D_{1/2}$ is the representation of A_1 in its original form $\mathfrak{sl}(2, \mathbb{C})$. D_1 is the adjoint representation (see the example in §1.5, with X_+, H, X_- as v_0, v_1, v_2).

There is a simple and concrete model for all the D_s (as reps of $\mathfrak{sl}(2, \mathbb{C})$, and also of the group $SL(2, \mathbb{C})$), starting with $D_{1/2}$ as the original action on $\mathbb{C}^2$. Namely, D_s is the induced rep in the space $S^{2s}\mathbb{C}^2$ of symmetric tensors of rank $2s$ (a subspace of the $2s$-fold tensor power of $\mathbb{C}^2$) or equivalently the $2s$-fold symmetric power of $\mathbb{C}^2$. Writing u and v for the two standard basis vectors $(1,0)$ and $(0,1)$ of $\mathbb{C}^2$, this is simply the space of the homogeneous polynomials of degree $2s$ in two symbols u and v. Here the element $g = \begin{bmatrix} a & b \\ c & d \end{bmatrix}$ of $SL(2, \mathbb{C})$ acts through the substitution $u \to au + cv, v \to bu + dv$, and the element $X = \begin{bmatrix} \alpha & \beta \\ \gamma & -\alpha \end{bmatrix}$ of $\mathfrak{sl}(2, \mathbb{C})$ acts through the *derivation* (i.e.,$X(p \cdot q) = Xp \cdot q + p \cdot Xq$) with $Xu = \alpha u + \gamma v, Xv = \beta u - \alpha v$. This action of $\mathfrak{sl}(2, \mathbb{C})$ can be described with standard differential operators: H acts as $u\partial_u - v\partial_v$, X_+ as $u\partial_v$, and X_- as $v\partial_u$. To show that this is indeed the promised rep, one verifies that these differential operators satisfy the commutation relations of H, X_+, and X_- (so that we have a rep), that the largest eigenvalue of H is $2s$ (operating on u^{2s}), and that the dimension of the space is correct, namely $2s+1$.

(Warning: The u and v are not the components of the vectors of $\mathbb{C}^2$. These components, say x and y, undergo the transformation usually written as

$$\begin{bmatrix} x \\ y \end{bmatrix} \to \begin{bmatrix} a & b \\ c & d \end{bmatrix} \begin{bmatrix} x \\ y \end{bmatrix} \text{ resp } \begin{bmatrix} \alpha & \beta \\ \gamma & -\alpha \end{bmatrix} \begin{bmatrix} x \\ y \end{bmatrix},$$

i.e., $x \to ax + by, y \to cx + dy$, resp $x \to \alpha x + \beta y$, $y \to \gamma x - \alpha y$. With x and y interpreted (as they should be) as the dual basis of the dual space to $\mathbb{C}^2$, this describes the transposed action of the original one, with the transposed matrix. Thus we are in the wrong space (although it is quite naturally isomorphic to $\mathbb{C}^2$) and we don't have a representation (but an anti-representation). The second trouble can be remedied by using the inverse, resp negative, and thus getting the contragredient representation $D_s^{\vartriangle}$. And it so happens that D_s is equivalent to its dual (it is self-contragredient, see §3.9), so that the trouble is not serious.)

There is another classical model for the D_s with integral s, which is of interest; we describe it briefly. As noted above, we may take $\mathfrak{o}(3, \mathbb{C})$ instead of $\mathfrak{sl}(2, \mathbb{C})$ or, even simpler, the real Lie algebra $\mathfrak{o}(3)$.

We write $\mathbb{R}^3$ with the three coordinates x, y, z, and consider the (infinite-dimensional) vectorspace P of polynomials in x, y, z with complex coefficients. There is a natural induced action of $\mathfrak{o}(3)$ on this space (and more generally on the space of all complex-valued C^∞-functions) as *differential operators*: R_x, R_y, R_z act, respectively, as

$$\begin{aligned} L_x &= z\partial_y - y\partial_z \\ L_y &= x\partial_z - z\partial_x \\ L_z &= y\partial_x - x\partial_y. \end{aligned}$$

(Verify that the L's satisfy the correct commutation relations. Physicists like to take instead the operators $J_x = i \cdot L_x$, etc., the *angular momentum operators*, because these versions are self-adjoint wr to the usual inner product between complex-valued functions.) There is also the *Laplace operator* $\Delta = \partial_x^2 + \partial_y^2 + \partial_z^2$; the polynomials (or functions) annulled by it are the *harmonic* ones. Δ commutes with the L's. It is easy to see (e.g., using the coordinates $w, \bar{w}, z$ defined below) that Δ maps the space P_s of polynomials of degree s onto P_{s-2}; one computes then the dimension of the harmonic subspace V_s of P_s as $2s + 1$.

Now to our representations: The harmonic space V_s is invariant under the L's. We claim that the induced representation is exactly D_s. To establish this , we note that the operator corresponding to the H of $\mathfrak{sl}(2, \mathbb{C})$ is $-2iR_z$ (see above); thus we have to find the eigenvalues of $-2iL_z$. To this end we introduce the new variables $w = x + iy$ and $\bar{w} = x - iy$, so that we write our polynomials as polynomials in $w, \bar{w}, z$. There are the usual operators $\partial_w = 1/2(\partial_x - i\partial_y)$ and $\partial_{\bar{w}} = 1/2(\partial_x + i\partial_y)$ with $\partial_w w = \partial_{\bar{w}}\bar{w} = 1, \partial_w \bar{w} =$

$\partial_{\bar{w}} w = 0$. They commute with each other, and we have $\Delta = 4\partial_w \partial_{\bar{w}} + \partial_z^2$ and $H = 2(\bar{w}\partial_{\bar{w}} - w\partial_w)$. We see that $w^a \cdot \bar{w}^b \cdot z^c$ is eigenvector of H with eigenvalue $2(b-a)$. In V^s the maximal eigenvalue of H is $2s$; it occurs only once, for the element $\bar{w}^s$, which happens to be harmonic. Thus the harmonic subspace has the dimension and the maximal eigenvalue of D_s and is therefore equivalent to it. $\surd$

For $s = 0$ the harmonic polynomials are just the constants, for $s = 1$ we have x, y and z, for $s = 2$ we find yz, zx, and xy and the $ax^2 + by^2 + cz^2$ with $a + b + c = 0$, a five-dimensional space.

The restrictions of the harmonic polynomials (as functions on $\mathbb{R}^3$) to the *unit sphere* (defined by $x^2 + y^2 + z^2 = 1$) are the classical *spherical harmonics*. Also note that our operators above are real and we could have worked with real polynomials; i.e., the real spherical harmonics are a real form of the space of spherical harmonics, in each degree.

1.12 Complete reduction for A_1

We prove the *complete reduction* theorem:

THEOREM A. *Every representation of A_1 is direct sum of irreducible ones (i.e., of D_s's).*

We give a special, pedestrian, proof, although later (§3.4) we shall bring a general and shorter proof. Our method is a modification of Casimir and van der Waerden's original one [4]. (This paper introduced what is now known as the *Casimir operator*, which turned out to be a very important object. In particular it leads to the simple general proof for complete reducibility alluded to above. It is interesting to note that Casimir and van der Waerden used the Casimir operator only for certain cases; the major part of their paper uses arguments of the kind described below.)
First we consider a representation on a vector space V with an invariant irreducible subspace V' and irreducible induced action on the quotient $W = V/V'$. (The general case will be reduced to this one by a simple argument.) We write $\pi : V \to W$ for the (equivariant) projection. Let the representation in V' be D_s, with basis $v_0, v_1, \ldots, v_r$ as in §1.11 (here $r = 2s$) and let the representation in W be D_q, with basis $w_0, w_1, \ldots, w_p$ (and $p = 2q$). We must produce an invariant complement U to V' in V.

The eigenvalues of H on V (with multiplicities) are those of D_s together with those of D_q. But it is not clear that H is diagonizable. In fact, that is the main problem. There are two cases.

(1) The easy case $q > s$, or $2s$ and $2q$ of different parity. Let u_0 be an eigenvector of H in V with eigenvalue $2q$. Clearly u_0 is not in V'. By

Lemma A of §1.11 we have $X_+u_0 = 0$, since $2q+2$ is not eigenvalue of H. But then, as described in §1.11, u_0 generates an invariant subspace U of type D_q, which is obviously complementary to V'.

(2) $q \leq s$, and $2s$ and $2q$ have the same parity. Put $d = 2e = r - p$, and note $Hv_e = 2qv_e$, by $r - 2e = 2q$. We show first that there is another eigenvector of H to this eigenvalue.

If not, then there exists a vector u_0, not in V', with $Hu_0 = 2qu_0 + v_e$ (namely a vector annulled by $(H - 2q)^2$, but not by $H - 2q$ itself); we may arrange $\pi(u_0) = w_0$. We form $u_1 = X_-u_0, u_2 = X_-u_1, \ldots$ and prove inductively (using $HX_- = X_-H - 2X_-$) the relation $Hu_i = (2q-2i)u_i + v_{e+i}$. We now distinguish the cases $q < s$ and $q = s$.

(a) If $q < s$, then u_{p+1} lies in V', since by equivariance we have $\pi(u_{p+1}) = X_-w_p = 0$. But no v in V' can satisfy the relation $Hv = (2q - 2p - 2)v + v_{e+p+1}$ (write v as $\Sigma a_i v_i$ and apply the diagonal matrix H). So this case cannot occur.

(b) If $q = s$ (i.e., $e = 0$), we find $Hu_{r+1} = (-2s - 2)u_{r+1}$, since v_{r+1} is 0; this implies $u_{r+1} = 0$, since $-2s - 2$ is not eigenvalue of H on V. We prove now, by induction, the formula $X_+u_i = \mu_i u_{i-1} + iv_{i-1}$ (with μ_i as in §1.11): First $X_+u_0 = 0$. This follows from $HX_+u_0 = X_+Hu_0 + 2X_+u_0 = (2s+2)X_+u_0$ (because $X_+v_0 = 0$); but $2s+2$ is not eigenvalue of H. Next, $X_+u_1 = X_+X_-u_0 = X_-X_+u_0 + Hu_0 = 2su_0 + v_0$, etc. (For the factor μ_i note $X_+u_i \equiv \mu_i u_{i-1} \bmod V'$, by applying π.) For $i = r+1$ we now get a contradiction, since u_{r+1} and μ_{r+1} vanish, but v_r does not.

Thus H has a second eigenvector to eigenvalue $2q$, in addition to v_e. In fact there is such a vector, u_0, that also satisfies $X_+u_0 = 0$. This is automatic if $q = s$; in the case $q < s$ it follows from Lemma A in §1.11, since the eigenvalue $2q+2$ of H has multiplicity 1. And now the vector u_0 generates the complementary subspace U that we were looking for. $\checkmark$

We come now to the general case. Let A_1 act on V, and let V_1 be an irreducible invariant subspace (which exists by the minimal dimension argument); let again π be the quotient map of V onto $W = V/V_1$. By induction over the dimension we may assume the action of A_1 on W completely reducible, so that W is direct sum of irreducible invariant subspaces W_i, with $i = 2, \ldots, k$. Put $W_i' = \pi^{-1}(W_i)$. We have the exact sequences $0 \to V_1 \to W_i' \to W_i \to 0$, with irreducible subspace and quotient. As proved above, there exists an invariant (and irreducible) complement V_i to V_1 in W_i'. It is easy to see now that V is direct sum of the V_i with $i = 1, \ldots, k$; complete reduction is established. $\checkmark$

The number of times a given D_s appears in the complete reduction of a representation φ is called the *multiplicity* n_s of D_s in φ. One writes $\varphi = \sum n_s D_s$. (Of course usually one lists only the—finitely many—non-zero n_s's.) The whole decomposition (i.e., the n_s's) is determined by the eigenvalues of H (and their multiplicities). For instance, the n_s for the largest s is equal to the multiplicity of the largest eigenvalue of H.

In particular, it is quite easy to work out the decomposition of the tensor product of two D_s's (the definition $H(v\otimes w) = Hv\otimes w + v\otimes Hw$ shows that the eigenvalues of H are the sums of the eigenvalues for the two factors). The result is

$$D_s \otimes D_t = D_{s+t} + D_{s+t-1} + D_{s+t-2} + \cdots + D_{|s-t|} \, .$$

(Verify that the eigenvalues of H, including multiplicities, are the same on the two sides of the formula.)

This relation is known as the *Clebsch-Gordan series*; it plays a role in quantum theory (angular momentum, spin, ...).

We add two more remarks about the D_s, namely about invariant bilinear forms and about invariant anti-involutions on their carrier spaces.

As noted earlier, $\mathfrak{sl}(2,\mathbb{C})$ is also $\mathfrak{sp}(1,\mathbb{C})$ – there is the invariant skew-symmetric form $x_1y_2 - x_2y_1$ or $\det[XY]$ on $\mathbb{C}^2$. This form induces invariant bilinear forms q_s on the symmetric powers of $\mathbb{C}^2$, i.e., on the carrier spaces of the D_s. For half-integral s (even dimension $2s+1$) the form turns out skew-symmetric, and so D_s is symplectic (meaning that all the operators are in the symplectic Lie algebra wr to q_s). For integral s (odd dimension $2s+1$) the form turns out symmetric, and so D_s is orthogonal (all operators are in the orthogonal Lie algebra of q_s). Explicitly this looks as follows:

For the representation space of D_s we take the physicists' basis $\{v_m\}$ with $m = -s, -s+1, -s+2, \ldots, s-1, s$. Then q_s is given by $q_s(v_m, v_{-m}) = (-1)^{s-m}$ and by $q_s(v_i, v_j) = 0$ if $i \neq -j$. (This is skew for half-integral s and symmetric for integral s.) Invariance under H is clear, since v_m and v_{-m} are eigenvectors with eigenvalues $2m$ and $-2m$. Invariance under X_+ and X_- takes a little more computation.

Now to the second topic: An *anti-involution* on a complex vector space V is a conjugation (an $\mathbb{R}$– linear operator on V (i.e.,on $V_{\mathbb{R}}$), say σ, with $\sigma(iv) = -i\sigma(v)$) that satisfies the relation $\sigma \circ \sigma = \pm id$.

In the case $+id$ (*first kind*) the eigenvalues of σ are ± 1. Let V_+, resp V_-, be the $+1$-, resp -1-, eigenspace of σ. Then V_- is $i \cdot V_+$ and $V_{\mathbb{R}}$ is the direct sum of V_+ and V_-.

In the case $-id$ (*second kind*) one can make V into a *quaternionic* vectorspace, by defining multiplying by the quaternion unit j as applying σ. (Usually one lets the quaternions act on V from the right side.)

On $\mathbb{C}^2$ there is a familiar anti-involution, of the second kind, say σ, namely "going to the unitary perpendicular": In terms of the basis $\{u, v\}$ defined earlier we have $\sigma(u) = v$ and $\sigma(v) = -u$, and generally $\sigma(au+bv) = -\bar{b}u + \bar{a}v$.

Next we recall that, as noted at the beginning of this section, in $\mathfrak{sl}(2,\mathbb{C})$ we find the real sub Lie algebra $\mathfrak{su}(2)$. It is geometrically clear, and easily verified by computation, that σ commutes with the elements of $\mathfrak{su}(2)$. Thus according to what we said above, we can regard $\mathbb{C}^2$ as (one-dimensional) quaternion space, and the action of $\mathfrak{su}(2)$ is quaternion-linear.

This extends in the obvious way to the other D_s: as described earlier, the carrier spaces are spaces of homogeneous polynomials (of degree $2s$) in u and v, and so σ induces anti-involutions in them. These are of the first, resp second, kind when s is integral, resp half-integral. Of course σ still commutes with the action of $\mathfrak{su}(2)$ (via D_s).

Thus for half-integral s we have quaternionic spaces, on which $\mathfrak{su}(2)$ acts quaternion-linearly.

For integral s the rep D_s (restricted to $\mathfrak{su}(2)$) is real in the sense that the $+1$-eigenspace of σ is a real form of the carrier space, invariant under the operators of $\mathfrak{su}(2)$. (Thus in a suitable coordinate system all the representing matrices will be real.) It also turns out that the form q_s is positive definite there. All this becomes clearer if we remember that $\mathfrak{su}(2)$ is isomorphic to $\mathfrak{o}(3)$. So we found that the D_s for integral s, as representation spaces of $\mathfrak{o}(3)$, are real; but we know that already from our discussion of the spherical harmonics. In particular $D_1|\mathfrak{o}(3)$ is the representation of $\mathfrak{o}(3)$ "by itself" on $\mathbb{R}^3$, with q_1 corresponding to $x^2+y^2+z^2$.

We discuss these matters in greater generality in §3.10.

2

Structure Theory

In this chapter we develop the structure theory of the general semisimple Lie algebra over $\mathbb{C}$ (the *Weyl-Chevalley normal form*) and bring the complete *classification of semisimple Lie algebras* (after W. Killing and E. Cartan). — Throughout $\mathfrak{g}$ is a complex Lie algebra, of dimension n, semisimple from §2.3 on. The concepts from linear algebra employed are described briefly in the Appendix.

2.1 Cartan subalgebra

A *Cartan sub Lie algebra* (commonly called *Cartan subalgebra*, *CSA* in brief, and usually denoted by $\mathfrak{h}$) is a nilpotent sub Lie algebra that is equal to its own normalizer in $\mathfrak{g}$. This somewhat opaque definition is the most efficient one. We will see later that for semisimple Lie algebras it is equivalent to $\mathfrak{h}$ being maximal Abelian with $\operatorname{ad} H$ semisimple (diagonizable) for all H in $\mathfrak{h}$. (Remark, with apologies: the arbitrary H in $\mathfrak{h}$ that appears here and will appear frequently from here on has to be distinguished from the specific element H of $\mathfrak{sl}(2, \mathbb{C})$ (see §1.1).) For $\mathfrak{gl}(n, \mathbb{C})$ a *CSA* is the set of all diagonal matrices—clearly an object of interest.

We write l for the dimension of $\mathfrak{h}$; this is called the *rank* of $\mathfrak{g}$, and we shall see later that it does not depend on the choice of $\mathfrak{h}$.

We establish existence and develop the important properties:

Let X be an element of $\mathfrak{g}$. Then $\operatorname{ad} X$ is an operator on the vector space $\mathfrak{g}$, and so there is the primary decomposition $\mathfrak{g} = \bigcup_\lambda \mathfrak{g}_\lambda(X)$, where λ runs through the eigenvalues of $\operatorname{ad} X$ and $\mathfrak{g}_\lambda(X)$ is the nilspace of $\operatorname{ad} X - \lambda$. (We recall that $\mathfrak{g}_\lambda(X)$ consists of all elements Y of $\mathfrak{g}$ that are nullified by some power of $\operatorname{ad} X - \lambda$. This makes sense for any λ, but is different from 0 only if λ is an eigenvalue of $\operatorname{ad} X$.)

The special nature of the operators in $\operatorname{ad} \mathfrak{g}$ finds its expression in the relations

(1) $$[\mathfrak{g}_\lambda(X), \mathfrak{g}_\mu(X)] \subset \mathfrak{g}_{\lambda+\mu}(X).$$

(The right-hand side is 0, if $\lambda + \mu$ is not eigenvalue of $\operatorname{ad} X$; i.e., in that case $\mathfrak{g}_\lambda(X)$ and $\mathfrak{g}_\mu(X)$ commute.)

They follow from the identity $(\operatorname{ad} X - (\lambda + \mu)) \cdot [YZ] = [(\operatorname{ad} X - \lambda) \cdot Y, Z] + [Y, (\operatorname{ad} X - \mu) \cdot Z]$ (the Jacobi identity) and the expression for $(\operatorname{ad} X - (\lambda + \mu))^r \cdot [YZ]$ that results by iteration. In particular, $\mathfrak{g}_0(X)$ is a sub Lie algebra; it contains X, by $\operatorname{ad} X \cdot X = [XX] = 0$.

The element X is *regular* if the nility of $\operatorname{ad} X$ (the algebraic multiplicity of 0 as eigenvalue) is as small as possible (compared with all other elements of $\mathfrak{g}$); and *singular* in the contrary case. For any X in $\mathfrak{g}$, the coefficients of the characteristic polynomial $\det(\operatorname{ad} X - t) = (-1)^n(t^n - D_1(X)t^{n-1} + D_2(X)t^{n-2} - \ldots)$ are polynomial functions of X. Here $D_n(X), = \det X,$ is identically, since 0 is eigenvalue of $\operatorname{ad} X$, by $[XX] = 0$. Let $D_r(X)$ be the last (i.e., largest index) of the not identically zero coefficients. Then an element X is regular precisely if $D_r(X)$ is not 0. The regular elements form the algebraic set of zeros of D_r. (E.g., if $\mathfrak{g}$ is Abelian, all elements are regular.)

The next proposition shows that CSA's exist and gives a way to construct them.

PROPOSITION A. *If X is regular, then the sub Lie algebra $\mathfrak{g}_0(X)$ is a Cartan subalgebra.*

We first show nilpotence:

For any Y in $\mathfrak{g}_0(X)$ (in particular for X itself) we have $\operatorname{ad} Y.\mathfrak{g}_\lambda(X) \subset \mathfrak{g}_\lambda(X)$, by formula (1), for any λ. For a $\lambda \neq 0$ the operator $\operatorname{ad} X$, restricted to $\mathfrak{g}_\lambda(X)$, is non-singular (all eigenvalues of $\operatorname{ad} X$ on $\mathfrak{g}_\lambda(X)$ equal λ). By continuity there is a neighborhood U of X in $\mathfrak{g}_0(X)$ such that for any Y in U the restriction of $\operatorname{ad} Y$ to $\mathfrak{g}_\lambda(X)$ is also non-singular. It follows that the restriction of $\operatorname{ad} Y$ to $\mathfrak{g}_0(X)$ is nilpotent; otherwise the nility of $\operatorname{ad} Y$ would be smaller than that of $\operatorname{ad} X$. But then $\operatorname{ad} Y$ is nilpotent on $\mathfrak{g}_0(X)$ for all Y in $\mathfrak{g}_0(X)$ by "algebraic continuation": nilpotence amounts to the vanishing of certain polynomials (the entries of a certain power of the restriction of $\operatorname{ad} Y$ to $\mathfrak{g}_0(X)$); and if a polynomial vanishes on an open set, like U, it vanishes identically on $\mathfrak{g}_0(X)$. Engel's theorem now shows that $\mathfrak{g}_0(X)$ is a nilpotent Lie algebra.

Next we show that $\mathfrak{g}_0(X)$ is its own normalizer in $\mathfrak{g}$: $\operatorname{ad} X$ is non-singular on each $\mathfrak{g}_\lambda(X)$ with $\lambda \neq 0$; thus if $[XY], = \operatorname{ad} X.Y,$ belongs to $\mathfrak{g}_0(X)$, so must Y. $\surd$

We note: The results of the next two sections will imply that for *semisimple* $\mathfrak{g}$ a *CSA* can be defined as a sub Lie algebra that is maximal Abelian and has $\operatorname{ad} X$ semisimple (diagonizable) on $\mathfrak{g}$ for all its elements X.

2.2 Roots

Let $\mathfrak{h}$ be a CSA. Nilpotence implies that $\mathfrak{h}$ is contained in $\mathfrak{g}_0(H)$ (as defined in §1.1) for any element H of $\mathfrak{h}$. (For any H' in $\mathfrak{h}$ we have $(\operatorname{ad} H)^r.H' = [HH \ldots HH'] = 0$ for large r.) Thus, if H and H' are two elements of $\mathfrak{h}$, all $\mathfrak{g}_\lambda(H)$ are invariant under $\operatorname{ad} H'$, by formula (1) in §2.1, and it follows that each $\mathfrak{g}_\lambda(H)$ is direct sum of its intersections with the $\mathfrak{g}_\mu(H')$. (This is simply the primary decomposition of $\operatorname{ad} H'$ on $\mathfrak{g}_\lambda(H)$.) Furthermore, all these intersections are invariant under $\operatorname{ad}\mathfrak{h}$, again by (1) in §2.1.

Iterating this process with elements H'', H''',... of $\mathfrak{h}$ (we look for elements, under whose primary decomposition some subspace of the previous stage decomposes further; for dimension reasons we come to an end after a finite number of steps) we see that $\mathfrak{g}$ can be written as direct sum of subspaces invariant under $\operatorname{ad}\mathfrak{h}$ with the property that on each such subspace each operator $\operatorname{ad} H$, for any H in $\mathfrak{h}$, has only one eigenvalue. It follows from Lie's Theorem (§1.8) that for each of these subspaces the (unique) eigenvalue of $\operatorname{ad} H$, as function of H, is a linear function on $\mathfrak{h}$. (This is clear in the triangularized form of the action.)

As an example: The subspace $\mathfrak{g}_0$ corresponding to the linear function 0, i.e., the intersection of the nilspaces on $\mathfrak{g}$ of all $\operatorname{ad} H$ with H in $\mathfrak{h}$ (which contains $\mathfrak{h}$) is $\mathfrak{h}$ itself: Apply Lie's Theorem to the action of $\mathfrak{h}$ on the quotient $\mathfrak{g}_0/\mathfrak{h}$; all eigenvalues (= diagonal elements) are 0. If $\mathfrak{g}_0$ were different from $\mathfrak{h}$, one could then find a vector Y, not in $\mathfrak{h}$, with $[HY]$ in $\mathfrak{h}$ for all H in $\mathfrak{h}$; but $\mathfrak{h}$ is its own normalizer in $\mathfrak{g}$.

We restate all this as follows: For each linear function λ on $\mathfrak{h}$ (= element of the dual space $\mathfrak{h}^\top$) denote by $\mathfrak{g}_\lambda$ the intersection of the nilspaces of all the operators $\operatorname{ad} H - \lambda(H)$ on $\mathfrak{g}$, with H running over $\mathfrak{h}$. Those λ, different from 0, for which $\mathfrak{g}_\lambda$ is not 0, are called the *roots* of $\mathfrak{g}$ wr to $\mathfrak{h}$; there are only finitely many such, of course; they are usually denoted by $\alpha, \beta, \gamma, \ldots$. The subset of $\mathfrak{h}^\top$ formed by them is denoted by Δ. To each α in Δ there is a subspace $\mathfrak{g}_\alpha$ of $\mathfrak{g}$, invariant under $\operatorname{ad}\mathfrak{h}$, called the *root space* to α, such that

(a) $\mathfrak{g}$ is direct sum of $\mathfrak{h}$ and the $\mathfrak{g}_\alpha$, for α in Δ,

(b) for each α in Δ and each H in $\mathfrak{h}$ the operator $\operatorname{ad} H$ has only one eigenvalue on $\mathfrak{g}_\alpha$, namely $\alpha(H)$, the value of the linear function α on H.

(As a matter of fact, for each α all the $\operatorname{ad} H$ on $\mathfrak{g}_\alpha$ have a simultaneous triangularization, with $\alpha(H)$ on the diagonal, by Lie's Theorem.) Occasionally we write Δ_0 for $\Delta \cup 0$. We note that Δ is not a subgroup of $\mathfrak{h}^\top$: it is after all a finite subset; in general "α and β in Δ" neither implies nor excludes "$\alpha + \beta$ in Δ". Clearly (1) of §1.1 implies

(2) $\quad [\mathfrak{g}_\lambda, \mathfrak{g}_m] \subset \mathfrak{g}_{\lambda+\mu} \qquad (= 0, \text{if} \lambda + \mu \text{ is not in } \Delta_0)$ for all λ, μ in $\mathfrak{h}^\top$.

We recall the Killing form κ or $\langle , \rangle$ (see §1.5). We call two elements X, Y of $\mathfrak{g}$ *orthogonal* to each other, in symbols $X \perp Y$, if $\langle X, Y\rangle$ is 0. We have

(3) $$\mathfrak{g}_\lambda \perp \mathfrak{g}_\mu, \text{ unless } \lambda + \mu = 0, \text{ for all } \lambda, \mu \text{ in } \mathfrak{h}^\top.$$

Proof: By (2) we have $[\mathfrak{g}_\lambda[\mathfrak{g}_\mu \mathfrak{g}_\nu]] \subset \mathfrak{g}_{\lambda+\mu+\nu}$ for all ν in Δ; i.e., for X in $\mathfrak{g}_\lambda$ and Y in $\mathfrak{g}_\mu$ the operator $\operatorname{ad} X \cdot \operatorname{ad} Y$ sends $\mathfrak{g}_\nu$ into $\mathfrak{g}_{\lambda+\mu+\nu}$. Since $\mathfrak{g}$ is direct sum of the $\mathfrak{g}_\nu$ with ν in Δ_0, we see by iteration that $\operatorname{ad} X \cdot \operatorname{ad} Y$ is nilpotent, if $\lambda + \mu$ is not 0; and so $\langle X, Y\rangle = \operatorname{tr}(\operatorname{ad} X \cdot \operatorname{ad} Y) = 0$. $\surd$

In particular $\mathfrak{h}$ is orthogonal to all the rootspaces $\mathfrak{g}_\alpha$ for α in Δ, and κ is identically 0 on each $\mathfrak{g}_\alpha$.

Finally, since all the $\operatorname{ad} X$ on each $\mathfrak{g}_\alpha$ can be taken triangular, we have the explicit formula

(4) $$\kappa(H, H') = \sum n_\alpha \cdot \alpha(H) \cdot \alpha(H'), \text{ for } H, H' \text{ in } \mathfrak{h}, \text{ with } n_\alpha = \dim \mathfrak{g}_\alpha.$$

(For nilpotent $\mathfrak{g}$ we have $\mathfrak{h} = \mathfrak{g}$. For $\mathfrak{aff}(1)$ (see §1.1) we can take $\mathbb{C}X_2$ as *CSA*.)

2.3 Roots for semisimple $\mathfrak{g}$

From here for the rest of the chapter we take $\mathfrak{g}$ semisimple, so that the Killing form is non-degenerate. This has many consequences:

(a) If all roots vanish on an element H of $\mathfrak{h}$, then H is 0.

Proof: H is orthogonal to all Y in $\mathfrak{h}$, by (4) of §2.2. As noted after (3) in §2.2, H is orthogonal to all $\mathfrak{g}_\alpha$ for α in Δ. Thus $\langle H, Y\rangle = 0$ for all Y in $\mathfrak{g}$. Non-degeneracy now implies $H = 0$.

(b) $((\Delta)) = \mathfrak{h}^\top$. I.e., the roots span $\mathfrak{h}^\top$; there are l linearly independent roots.

This follows by vector space duality from (a).

(c) $\mathfrak{h}$ is Abelian.

Proof: $\operatorname{ad} Y$ on any $\mathfrak{g}_\alpha$ is (or can be taken) triangular for all Y in $\mathfrak{h}$. Then for H in $[\mathfrak{h}\mathfrak{h}]$ the eigenvalue on $\mathfrak{g}_\alpha$, i.e., the value $\alpha(H)$, is 0. Now (a) applies.

(d) The Killing form is non-degenerate on $\mathfrak{h}$.

This follows from the non-degeneracy on $\mathfrak{g}$ together with the fact that $\mathfrak{h}$ is orthogonal to all $\mathfrak{g}_\alpha$ for α in Δ (see (3) in §2.2).

(e) For every H in $\mathfrak{h}$ the operator $\operatorname{ad} H$ on $\mathfrak{g}$ is semisimple. Equivalently: For each root α we have $\operatorname{ad} H.X = \alpha(H) \cdot X$ for H in $\mathfrak{h}$ and X in $\mathfrak{g}_\alpha$.

(Put differently: $\operatorname{ad} H$ reduces on $\mathfrak{g}_\alpha$ to the scalar operator $\alpha(H)$.)

Proof: Let $\operatorname{ad} H = S + N$ be the Jordan decomposition. One shows first that S is a derivation of $\mathfrak{g}$: Namely S on $\mathfrak{g}_\alpha$ is multiplication by $\alpha(H)$. For X in $\mathfrak{g}_\alpha$ and Y in $\mathfrak{g}_\beta$, with α, β in Δ_0, we have $[SX, Y] + [X, SY] = [\alpha(H)X, Y] + [X, \beta(H)Y] = (\alpha+\beta)(H)[XY]$; and the latter is $S[XY]$ by (2) of §2.2. By §1.10 there is a Y in $\mathfrak{g}$ with $S = \operatorname{ad} Y$. Since $S \cdot Z = 0$ for all Z in $\mathfrak{h}$, Y is in the centralizer of $\mathfrak{h}$ and so actually in $\mathfrak{h}$. Also $\operatorname{ad}(H - Y), = N$, has only 0 as eigenvalue on $\mathfrak{g}$; i.e., all roots vanish on $H - Y$. By (a) we have $H - Y = 0$, and then also $N = 0$. $\surd$

(f) $\Delta = -\Delta$. I.e., if α is in Δ, so is $-\alpha$.

Proof: By (3) of §2.2 all $\mathfrak{g}_\beta$, except possibly $\mathfrak{g}_{-\alpha}$, and also $\mathfrak{h}$ are orthogonal to $\mathfrak{g}_\alpha$. By non-degeneracy of κ the space $\mathfrak{g}_{-\alpha}$ cannot be 0. $\surd$

2.4 Strings

PROPOSITION A. *For each α in Δ the subspace $[\mathfrak{g}_\alpha, \mathfrak{g}_{-\alpha}]$ of $\mathfrak{h}$ has dimension 1, and the restriction of α to it is not identically 0.*

Proof: For any X in $\mathfrak{g}_\alpha$ the operator $\operatorname{ad} X$ is nilpotent on $\mathfrak{g}$, since by (2) of §2.2 it maps $\mathfrak{g}_\beta$ to $\mathfrak{g}_{\beta+\alpha}$ (here, as often, $\mathfrak{g}_0$ means $\mathfrak{h}$); iterating one eventually gets to 0. If Y is in $\mathfrak{g}_{-\alpha}$ and $[XY]$ is 0, then $\operatorname{ad} X \cdot \operatorname{ad} Y$ is nilpotent (since then $\operatorname{ad} X$ and $\operatorname{ad} Y$ commute), and so $\langle X, Y\rangle$ vanishes. By (3) of §2.2 and non-degeneracy of κ there exist X_0 in $\mathfrak{g}_\alpha$ and Y_0 in $\mathfrak{g}_\alpha$ with $\langle X_0, Y_0\rangle \neq 0$, and thus also with $[X_0 Y_0] \neq 0$. So $\dim[\mathfrak{g}_\alpha, \mathfrak{g}_{-\alpha}] > 0$.

For the remainder of the proof we need an important definition: For α and β in Δ the *$\alpha-$string of β* is (ambiguously) either the set of those forms $\beta + t\alpha$ with integral t that are roots or 0, or the direct sum, over all integral t, of the spaces $\mathfrak{g}_{\alpha+t\beta}$. We denote the string by $\mathfrak{g}_\beta^\alpha$; of course only the $\beta + t\alpha$ that are roots or 0 actually appear). (Actually 0 occurs only if β equals $-\alpha$ (see §2.5); and in that case we modify the definition of string slightly at the end of this section.)

By (2) of §2.2 clearly $\mathfrak{g}_\beta^\alpha$ is invariant under $\operatorname{ad} X$, resp $\operatorname{ad} Y$ for X in $\mathfrak{g}_\alpha$, resp Y in $\mathfrak{g}_\alpha$. It follows that for such X and Y the trace of $\operatorname{ad}[XY]$ (i.e., of $\operatorname{ad} X \cdot \operatorname{ad} Y - \operatorname{ad} Y \cdot \operatorname{ad} X$) on $\mathfrak{g}_\beta^\alpha$ is 0. Now for Z in $\mathfrak{h}$ the trace of $\operatorname{ad} Z$ on $\mathfrak{g}_\gamma$ is $n_\gamma \cdot \gamma(Z)$ (see (4) in §2.2 for n_γ), and so the trace on $\mathfrak{g}_\beta^\alpha$ is of the form $p\beta(Z) + q\alpha(Z)$ with $p = \dim \mathfrak{g}_\beta^\alpha$ and q integral. Taking $[XY]$ (which is in $\mathfrak{h}$ by (2) of §2.2) as Z, we see: if $\alpha([XY])$ is 0, so is $\beta([XY])$ for all β in Δ; but then $[XY]$ is 0 by (a) in §2.3. In other words, the intersection of $[\mathfrak{g}_\alpha, \mathfrak{g}_{-\alpha}]$ and of the nullspace of α is 0. Clearly this establishes the proposition. $\surd$

Since κ is non-degenerate on $\mathfrak{h}$, we have the usual isomorphism of $\mathfrak{h}$ with its dual; i.e., to each λ in $\mathfrak{h}^\top$ (in particular to each root) there is a unique element h_λ in $\mathfrak{h}$ with $\langle h_\lambda, Z\rangle = \lambda(Z)$ for all Z in $\mathfrak{h}$. The h_α for α in Δ (called *root vectors*) span $\mathfrak{h}$, by (b) of §2.3. We claim: h_α is an element of $[\mathfrak{g}_\alpha, \mathfrak{g}_{-\alpha}]$.

PROPOSITION B. *For X in $\mathfrak{g}_\alpha$ and Y in $\mathfrak{g}_{-\alpha}$ with $\langle X,Y\rangle = 1$ the element $[XY]$ equals h_α.*

Proof: $\langle [XY], Z\rangle = -\langle Y, [XZ]\rangle = \langle Y, [ZX]\rangle = \langle Y, \alpha(Z)X\rangle = \alpha(Z)$. Here the first $=$ comes from invariance of the Killing form, and the third from (e) in §2.3.

By Prop. A we have $\langle h_\alpha, h_\alpha\rangle = \alpha(h_\alpha) \neq 0$ (since h_α is of course not 0). We introduce the important elements $H_\alpha = (2/\langle h_\alpha, h_\alpha\rangle)\cdot h_\alpha$, for α in Δ; they are the *coroots* (of $\mathfrak{g}$ wr to $\mathfrak{h}$) and will play a considerable role. They span $\mathfrak{h}$ (just like the h_α) and satisfy the relations $\alpha(H_\alpha) = 2$ and $[\mathfrak{g}_\alpha, \mathfrak{g}_{-\alpha}] = \mathbb{C}h_\alpha (= \mathbb{C}\mathfrak{h}_{-\alpha})$. More is true about the $\mathfrak{g}_\alpha$.

PROPOSITION C. *For each α in Δ the dimension of $\mathfrak{g}_\alpha$ is 1, and $\mathfrak{g}_{t\alpha}$ is 0 for $t = 2, 3, \ldots$ (i.e., the multiples 2α, $3\alpha, \ldots$ are not roots).*

Proof: By Prop. B there exist root elements X_α in $\mathfrak{g}_\alpha$ and $X_{-\alpha}$ in $\mathfrak{g}_{-\alpha}$ so that $[X_\alpha X_{-\alpha}] = H_\alpha$. Using (e) of §2.3 and $\alpha(H_\alpha) = 2$, we see $[H_\alpha X_\alpha] = 2X_\alpha$ and $[H_\alpha X_{-\alpha}] = -2X_{-\alpha}$. Let $\mathfrak{q}_\alpha$ be the subspace of $\mathfrak{g}$ spanned by $X_{-\alpha}$, H_α, and all the $\mathfrak{g}_{t\alpha}$ for $t = 1, 2, 3\ldots$.

Proposition A of §2.4, (e) of §2.3 and (2) of §2.2 imply that $\mathfrak{q}_\alpha$ is invariant under the three operators X_α, $X_{-\alpha}$ and H_α. It follows from $\operatorname{ad} H_\alpha = \operatorname{ad}[X_\alpha, X_{-\alpha}] = [\operatorname{ad} X_\alpha, \operatorname{ad} X_{-\alpha}]$ that the trace of H_α on $\mathfrak{q}_\alpha$ is 0. From the scalar nature of H_α on $\mathfrak{g}_\alpha$ we see that this trace is $2(-1+n_\alpha + 2n_{2\alpha} + 3n_{3\alpha} + \ldots)$ (recall $n_\beta = \dim g_\beta$). Therefore we must have $n_\alpha = 1$ (and so $\mathfrak{g}_\alpha = \mathbb{C}X_\alpha$) and $n_{2\alpha} = n_{3\alpha} = \ldots = 0$.

We modify the definition of the α-string of β for the case $\beta = -\alpha$ by putting $\mathfrak{g}^{\pm\alpha}_{\pm\alpha} = ((X_\alpha, H_\alpha, X_{-\alpha}))$.

Note that (4) of §2.2 now becomes

$$(4') \qquad \langle X, Y\rangle = \sum \alpha(X)\cdot\alpha(Y) \text{ for all } X, Y \text{ in } \mathfrak{h}$$

2.5 Cartan integers

The bracket relations between the $H_\alpha, X_\alpha, X_{-\alpha}$ introduced above show that these three elements form a sub Lie algebra of $\mathfrak{g}$; we shall denote it by $\mathfrak{g}^{(\alpha)}$. (Note $H_{-\alpha} = -H_\alpha$ and $\mathfrak{g}^{(-\alpha)} = \mathfrak{g}^{(\alpha)}$). Quite clearly $\mathfrak{g}^{(\alpha)}$ is isomorphic to the Lie algebra A_1 that we studied in the last chapter, with H_α, X_α, and $X_{-\alpha}$ corresponding, in turn, to H, X_+, and X_-. This has important consequences. Namely from the representation theory of A_1 (§1.11) we know that in any representation the eigenvalues of H are integers, and are made up of sequences that go in steps of 2 from a maximum $+r$ to a minimum $-r$, one such sequence for each irreducible constituent. Now in the proof for Proposition A of §2.4 we saw in effect that the string $\mathfrak{g}_\beta^\alpha$ is $\mathfrak{g}^{(\alpha)}$-stable (even for the modified definition in case $\beta = -\alpha$); thus we have a representation there. The eigenvalues of $\operatorname{ad} H_\alpha$ on $\mathfrak{g}_\beta^\alpha$ are precisely the values $\beta(H_\alpha) + 2t$, for those integers t for which $\beta + t\alpha$ is a root or 0 (recall $\alpha(H_\alpha) = 2$), and the multiplicities are 1 (we have $\dim \mathfrak{g}_{\beta+t\alpha} = 1$ by Prop.B, §2.4). It is clear then that $\mathfrak{g}_\beta^\alpha$ is irreducible under $\mathfrak{g}^{(\alpha)}$ and the representation is one of the D_s's; in particular, these t-values occupy exactly some interval in $\mathbb{Z}$ (one describes this by saying that strings are unbroken). We have:

PROPOSITION A. *The values $\beta(H_\alpha)$, for α and β in Δ, are integers (they are denoted by $a_{\beta\alpha}$ and called the Cartan integers of $\mathfrak{g}$). For any α and β there are two non-negative integers p $(= p(\alpha, \beta))$ and q $(= q(\alpha, \beta))$, such that the $\mathfrak{g}_{\beta+t\alpha}$ that occur in $\mathfrak{g}_\beta^\alpha$ (i.e., that are not 0) are exactly those with $-q \le t \le p$.*

There is the relation

$$a_{\beta\alpha} = q - p. \tag{5}$$

(For $\beta = -\alpha$ the string consists of $\mathfrak{g}_{-\alpha}$, $((H_\alpha))$, and $\mathfrak{g}_\alpha$; and one has $a_{\alpha\alpha} = -a_{\alpha,-\alpha} = 2$.)

(In the literature one also finds the notation $a_{\alpha\beta}$ for the value $\beta(H_\alpha)$, instead of $a_{\beta\alpha}$.)

Relation (5) follows from the fact that the smallest eigenvalue, $\beta(H_\alpha) - 2q$, must be the negative of the largest one, $\beta(H_\alpha) + 2p$. (And the representation on $\mathfrak{g}_\beta^\alpha$ is the D_s with $2s = p+q$.) — We note that from the definition we have $a_{\beta\alpha} = \beta(H_\alpha) = 2\langle h_\beta, h_\alpha\rangle / \langle h_\alpha, h_\alpha\rangle$ and $a_{\alpha\alpha} = 2$.

$a_{\beta\alpha}$ can be different from $a_{\alpha\beta}$. We shall see soon that only the numbers $0, \pm 1, \pm 2, \pm 3$ can occur as $a_{\beta\alpha}$. We develop some more properties.

PROPOSITION B. *For any two roots α, β the combination $\beta - a_{\beta\alpha} \cdot \alpha$ is a root. In fact, with $\varepsilon = \operatorname{sign} a_{\beta\alpha}$ all the terms $\beta, \beta - \varepsilon\alpha, \beta - 2\varepsilon\alpha, \ldots, \beta - a_{\beta\alpha}\alpha$ are roots again (or 0).*

This follows from the fact that $a_{\beta\alpha}$ lies in the interval $[-p, q]$, by (5). (Here 0 can occur in the sequence only if $\beta = -\alpha$, by Prop.C.) We note a slightly different, very useful version.

PROPOSITION B′. *For two roots α and β with $\alpha \ne \beta$, if $\beta - \alpha$ is not a root (β is "α-minimal"), then one has $a_{\beta\alpha} \le 0$.*

Proof: q in (5) is now 0.

There is an important strengthening of Proposition B of §2.4.

PROPOSITION C. *A multiple $c \cdot \alpha$ of a root α with c in $\mathbb{C}$ is again a root iff $c = \pm 1$.*

Proof: The if-part is (f) in §2.3. For the only if, suppose $\beta = c\alpha$ is also a root. Evaluating on H_α and on H_β we get $a_{\beta\alpha} = 2c$ and $2 = c \cdot a_{\alpha\beta}$.Thus by Prop. A both $2c$ and $2/c$ are integers. It follows that c must be one of $\pm 1, \pm 2, \pm 1/2$. Prop.B of §2.4 forbids ± 2 and then also $\pm 1/2$. $\surd$

Generators X_α of $\mathfrak{g}_\alpha$, always subject to $[X_\alpha X_{-\alpha}] = H_\alpha$, will be called *root elements* (to be distinguished from the root vectors h_α of §2.4). One might say that $\mathfrak{g}$ is constructed by putting together a number of copies of A_1 (namely the $\mathfrak{g}^{(\alpha)}$), in such a way that the X_+'s and X_-'s are independent, but with relations between the H's [they all lie in $\mathfrak{h}$, and there are usually more than l ($= \dim \mathfrak{h}$) roots].

Integrality of the $a_{\beta\alpha}$ and formula (4′) of §2.4 imply that all inner products $\langle H_\alpha, H_\beta \rangle$ are integers.

2.6 Root systems, Weyl group

Let $\mathfrak{h}_0$ be the real subspace of $\mathfrak{h}$ formed by the real linear combinations of the H_α for α in Δ; we refer to $\mathfrak{h}_0$ as the *normal real form* of $\mathfrak{h}$. The values of the $\beta(H_\alpha)$ being integral, the roots of $\mathfrak{g}$ are (or better : restrict to) real linear functions on $\mathfrak{h}_0$.

PROPOSITION A. *The Killing form κ, restricted to $\mathfrak{h}_0$, is a (real) positive definite bilinear form .*

Proof: The Killing form is non-negative by (4′) of §2.4, and an equation $\langle X, X \rangle = 0$ implies that all $\alpha(X)$ vanish: this in turn implies $X = 0$, by (a) of §2.3. In the usual way, this defines the norm $|X| = \langle X, X \rangle^{1/2}$ on $\mathfrak{h}_0$.

The formula $\langle H_\alpha, H_\alpha \rangle = 4/\langle h_\alpha, h_\alpha \rangle$, easily established, shows that the $\langle h_\alpha, h_\alpha \rangle$, and then also all $\langle h_\alpha, h_\beta \rangle$ are rational numbers, so that the h_α are rational multiples of the H_α, and the h_α also span $\mathfrak{h}_0$. Furthermore:

PROPOSITION B. *$\mathfrak{h}_0$ is a real form of $\mathfrak{h}$.*

This means that any X in $\mathfrak{h}$ is uniquely of the form $X' + iX''$ with X' and X'' in $\mathfrak{h}_0$, or that $\mathfrak{h}_{\mathbb{R}}$ (i.e., $\mathfrak{h}$ with scalars restricted to $\mathbb{R}$) is the direct sum of $\mathfrak{h}_0$ and $i\mathfrak{h}_0$, or that any basis of $\mathfrak{h}_0$ over $\mathbb{R}$ is a basis of $\mathfrak{h}$ over $\mathbb{C}$

Proof: We have $\mathbb{C}\mathfrak{h}_0 = \mathfrak{h}$ (since the H_α span $\mathfrak{h}$), and so $\mathfrak{h}$ is at any rate spanned by $\mathfrak{h}_0$ and $i\mathfrak{h}_0$ (over $\mathbb{R}$). For any X in the intersection $\mathfrak{h}_0 \cap i\mathfrak{h}_0$ we have $X = iY$ with X and Y in $\mathfrak{h}_0$; therefore $0 \le \langle X, X\rangle$ (by positive definiteness) $= -\langle Y, Y\rangle$ (by $\mathbb{C}$-linearity of κ) ≤ 0 (positive definiteness again). So $\langle X, X\rangle = 0$ and then also $X = 0$. $\surd$

We consider the isomorphism of $\mathfrak{h}$ with its dual space $\mathfrak{h}^\top$, defined by the Killing form ($\lambda \leftrightarrow h_\lambda$ as in §2.4). Clearly the real subspace $\mathfrak{h}_0$ goes over into $((\Delta))_{\mathbb{R}}$, the $\mathbb{R}$-span of Δ, which we denote by $\mathfrak{h}_0^\top$; and clearly this is a real form of $\mathfrak{h}^\top$. We transfer the Killing form to $\mathfrak{h}^\top$ (and to $\mathfrak{h}_0^\top$) in the standard way, by putting $\langle \lambda, \mu\rangle = \langle h_\lambda, h_\mu\rangle$; the isomorphism (of $\mathfrak{h}$ with $\mathfrak{h}^\top$ and of $\mathfrak{h}_0$ with $\mathfrak{h}_0^\top$) is then an isometry. (E.g., the definition $a_{\beta\alpha} = \beta(H_\alpha)$ of the Cartan integers translates into $a_{\beta\alpha} = 2\langle\beta, \alpha\rangle/\langle\alpha, \alpha\rangle$.) It is fairly customary to identify $\mathfrak{h}$ and $\mathfrak{h}^\top$ under this map; however we prefer to keep space and dual space separate.

We collect some properties of Δ into an important definition. Let V be a Euclidean space, i.e., a vector space over $\mathbb{R}$ with a positive-definite inner product $\langle,\rangle$.

DEFINITION C. *An (abstract) root system (in V, wr to $\langle,\rangle$) is a finite non-empty subset, say R, of V, not containing* 0, *and satisfying*

(i) For α, β in R, $2\langle\beta, \alpha\rangle/\langle\alpha, \alpha\rangle$ is an integer (denoted by $a_{\beta\alpha}$)

(ii) For α, β in R, the vector $\beta - a_{\beta\alpha} \cdot \alpha$ is also in R,

(iii) If α and a multiple $r \cdot \alpha$ are both in R, then $r = \pm 1$.

(Strictly speaking this is a *reduced* root system; one gets the slightly more general notion of *unreduced* root system by dropping condition (iii). The argument for Proposition C in §2.5 shows that the additional r- values allowed then are ± 2 and $\pm 1/2$.)

Clearly the properties of the set Δ of roots of $\mathfrak{g}$ wr to $\mathfrak{h}$, developed above, show that it is a root system in $\mathfrak{h}_0^\top$.

Note that $a_{\alpha\alpha}$ equals 2, and that (i) and (ii) imply that $-\alpha$ belongs to R if α does. The *rank* of a root system R is the dimension of the subspace of V spanned by R. (Thus the rank of Δ equals the rank of $\mathfrak{g}$ as defined in §2.1.) We shall usually assume that R spans V.

Condition (ii) has a geometrical meaning: For any μ in V, $\neq 0$, let S_μ be the reflection of V wr to the hyperplane orthogonal to μ (this is an isometry of V with itself; it is the identity map on that hyperplane and sends μ into $-\mu$). It is a simple exercise to derive the formula

(7) $$S_\mu(\lambda) = \lambda - 2\langle\lambda, \mu\rangle/\langle\mu, \mu\rangle \cdot \mu, \text{ for all } \lambda \text{ in } V.$$

We see that condition (ii) can be restated as

(ii)′ *If α and β are in R, so is $S_\alpha(\beta)$. Equivalently, the set R is invariant under all S_α.*

Similarly, (i) can be restated as

(i)′ *The difference $S_\alpha(\beta) - \beta$ is an integral multiple of α.*

The S_α, for α in R, generate a group of isometries of V, called the *Weyl group* $\mathcal{W}$ of R (or of $\mathfrak{g}$, wr to $\mathfrak{h}$, if R is the set Δ of roots of $\mathfrak{g}$ wr to $\mathfrak{h}$). The S_α are the *Weyl reflections.*

Clearly any S in $\mathcal{W}$ leaves R invariant. It is also clear that each such S is completely determined by the permutation of the elements of R determined by it (and that S is the identity on the orthogonal complement of $((R))$ in V). This implies that $\mathcal{W}$ is a finite group.

Two root systems R_1 and R_2 are *equivalent*, if there exists a *similarity* (= isometry up to a constant factor) of $((R_1))$ onto $((R_2))$ that sends the set R_1 onto the set R_2. A root system is *simple*, if it is not union of two non-empty subsets that are orthogonal to each other, and *decomposable* in the opposite case. Obviously any root system is union of simple ones that are pairwise orthogonal, and the splitting is unique.

Conversely, given two root systems P and Q, there is a well-defined *direct sum* $P \oplus Q$, namely the union of P and Q in the direct sum of the associated vector spaces, with the usual inner product. We note that the set $\{h_\alpha\}$ of root vectors of $\mathfrak{g}$ wr to $\mathfrak{h}$ is a root system in $\mathfrak{h}_0$, equivalent and even isometric to the root system Δ of roots of $\mathfrak{g}$ wr to $\mathfrak{h}$, in $\mathfrak{h}^\top$.

We interpolate a simple geometric observation.

PROPOSITION D. *The Weyl group of a simple root system R acts irreducibly on the vector space V of R.*

In particular, the $\mathcal{W}$-orbit of any non-zero vector spans V.

Proof: A subspace W of V is stable under a reflection S_λ, for some λ in V, iff it is either orthogonal to λ or contains λ. Thus, if W is stable under the Weyl group, in particular under all the S_α for the α in R, it divides R into two sets: the α orthogonal to W and the α in W. By simplicity of R one of these two sets is empty, which implies that W is either 0 or V. $\surd$

With every root system $R = \{\alpha\}$ there is associated a *dual* or *reciprocal* root system $R' = \{\alpha'\}$ in the same vector space, defined by $\alpha' = 2/\langle\alpha,\alpha\rangle\cdot\alpha$. (Except for the factor 2 this comes from the "transformation by reciprocal radii": we have $|\alpha'| = 2 \cdot |\alpha|^{-1}$.) One computes $\langle\alpha',\alpha'\rangle = 4/\langle\alpha,\alpha\rangle$; and the Cartan integers of R' are related to those of R by $a_{\beta'\alpha'} = a_{\alpha\beta}$. Thus condition (i) holds. Condition (ii), in the form (ii)′, invariance of R under the Weyl reflections S_α, is also clear, once one notices $S_\alpha = S_{\alpha'}$ (i.e., $\mathcal{W} = \mathcal{W}'$). Condition (iii) is obvious. Thus R' is a root system. Clearly we have $R'' = R$.

The importance of the process of assigning to each semisimple Lie algebra $\mathfrak{g}$ the root system Δ of its roots wr to a Cartan sub Lie algebra lies in the following three facts (to be established in the rest of the chapter):

A. The whole Lie algebra $\mathfrak{g}$ (in particular the bracket operation) can be reconstructed from the root system Δ (Weyl-Chevalley normal form).

B. To each (abstract) root system there corresponds a semisimple Lie algebra.

C. The root systems are easily classified.

In other words: there is a bijection between the set of (isomorphism classes of) semisimple Lie algebras and the set of (equivalence classes of) root systems, and the latter set is easily described. That gives the *Cartan-Killing classification* of semisimple Lie algebras.

We begin with A.

2.7 Root systems of rank two

We determine all root systems of rank two (and also those of rank one), as examples, but mainly because they are needed for later constructions. Clearly there is only one root system of rank one; it consists of two non-zero vectors α and $-\alpha$; the Cartan integers are $a_{\alpha\alpha} = a_{-\alpha,-\alpha} = -a_{\alpha,-\alpha} = -a_{-\alpha,\alpha} = 2$. We denote this system by A_1. It is indeed the root system of the Lie algebra $A_1 (= \mathfrak{sl}(2, \mathbb{C}))$. Here $((H))$ is a CSA; the rank is 1; the equations $[H, X_\pm] = \pm 2X_\pm$ mean that there is a pair of roots $\alpha, -\alpha$ with $\pm\alpha(cH) = \pm 2c$; in particular, $\pm\alpha(H) = \pm 2$, so that $\pm H$ is the coroot to $\pm\alpha$, and the real form $\mathfrak{h}_0$ of the CSA is $\mathbb{R}H$.

Let now R be any root system, and consider two of its elements, α and β. From the definition we have $a_{\alpha\beta} \cdot a_{\beta\alpha} = 4\langle\alpha,\beta\rangle^2/|\alpha|^2|\beta|^2 = 4cos^2\theta$, where θ means the angle between α and β in the usual sense ($0 \le \theta \le \pi$). The $a_{\alpha\beta}$'s being integers, the possible values of $a_{\alpha\beta} \cdot a_{\beta\alpha}$ are then $0, 1, 2, 3, 4$ (this is a crucial point for the whole theory!). The value 4 means dependence of the two vectors ($\cos\theta = \pm 1$), and so $\alpha = \pm\beta$, by condition (iii) for root systems. For the discussion of the other cases we assume $a_{\beta\alpha} \le 0$ (i.e., $\theta \ge \pi/2$); for this we may have to replace β by $S_\alpha(\beta)$; it is easily seen that this just changes the signs of $a_{\alpha\beta}$ and $a_{\beta\alpha}$. The value 0 corresponds to α and β being orthogonal to each other ($\alpha \perp \beta, \theta = \pi/2$); or, equivalently, $a_{\alpha\beta} = a_{\beta\alpha} = 0$. For the remaining three cases integrality of the a's implies that one of the two is -1, and the other is -1 or -2 or -3; the corresponding angles θ are $2\pi/3, 3\pi/4, 5\pi/6$. In these three cases we also get $|\beta|^2/|\alpha|^2 = a_{\beta\alpha}/a_{\alpha\beta} = 1$ or 2 or 3 or their reciprocals (whereas in the case of 0 we get no restriction on the ratio of $|\alpha|$ and $|\beta|$). We see that there are very few possibilities for the "shape" of the pair α, β. We arrange the facts in a table and a figure, taking α to be the shorter of the two vectors:

Case	$a_{\alpha\beta}$	$a_{\beta\alpha}$	θ	$\lvert\beta\rvert/\lvert\alpha\rvert$
(i)	0	0	$\pi/2$	?
(ii)	-1	-1	$2\pi/3$	1
(iii)	-1	-2	$3\pi/4$	$\sqrt{2}$
(iv)	-1	-3	$5\pi/6$	$\sqrt{3}$

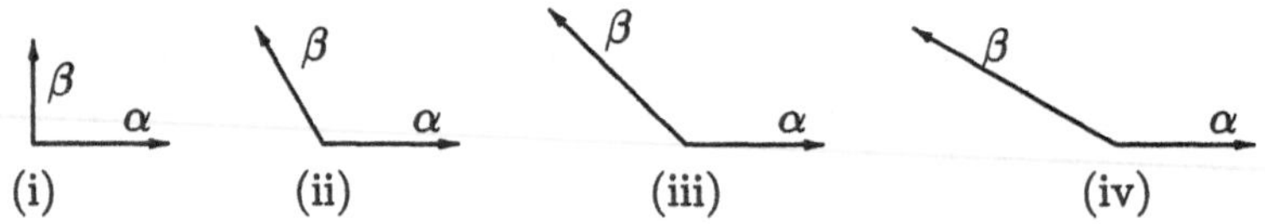

Figure 1

The change needed for the case $a_{\alpha\beta} \geq 0$ is the removal of all minus-signs and the replacement of θ by $\pi - \theta$; the only acute angles possible are $\pi/6, \pi/4, \pi/3$ (and $\pi/2$).

We come now to the root systems of rank 2.

PROPOSITION A. *Any root system of rank two is equivalent to one of the four shown in Figure 2 below:*

(i) $A_1 \oplus A_1$

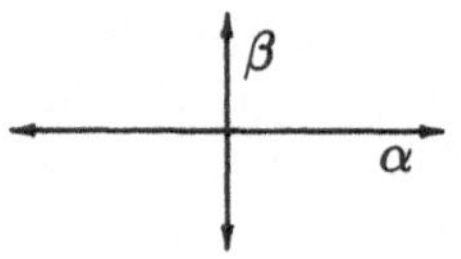

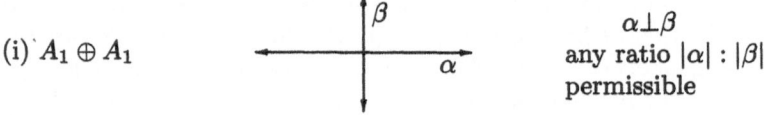

$\alpha \perp \beta$
any ratio $|\alpha| : |\beta|$ permissible

(ii) A_2

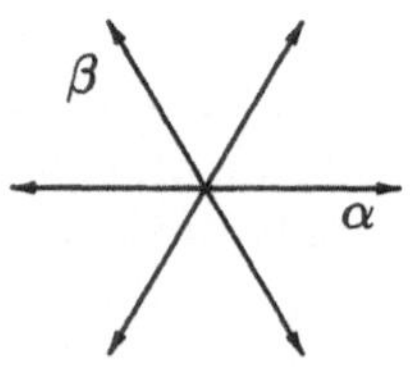

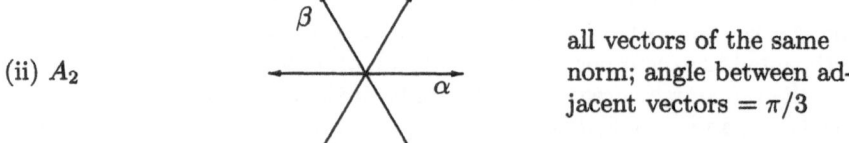

all vectors of the same norm; angle between adjacent vectors $= \pi/3$

(iii) B_2

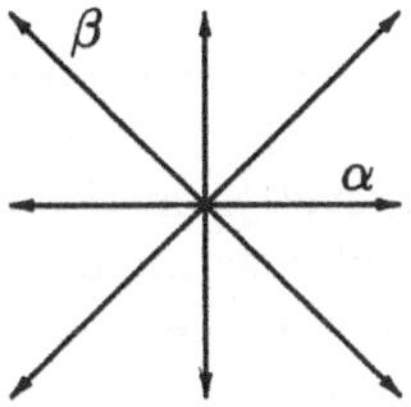

Ratio $|\beta| : |\alpha| = \sqrt{2}$; angle between adjacent vectors $= \pi/4$

(iv) G_2

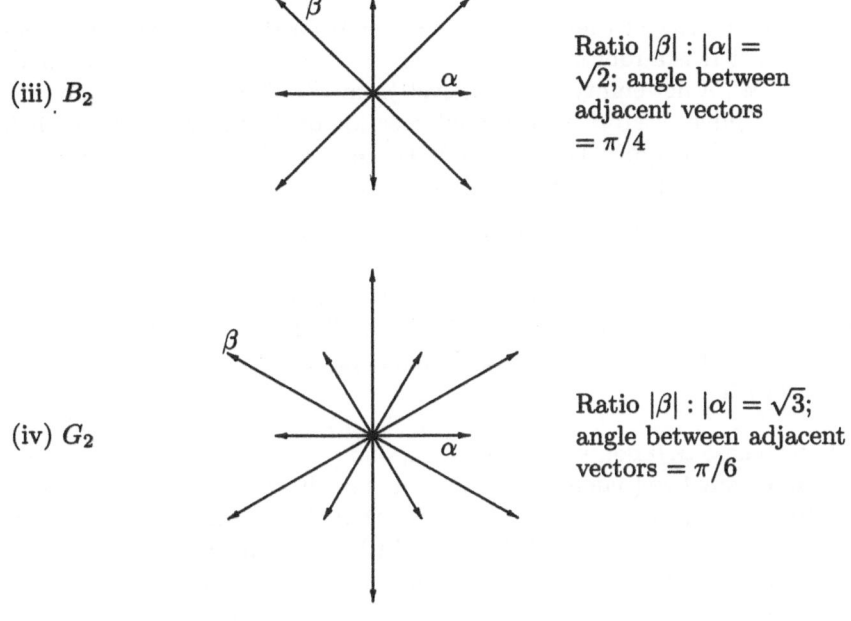

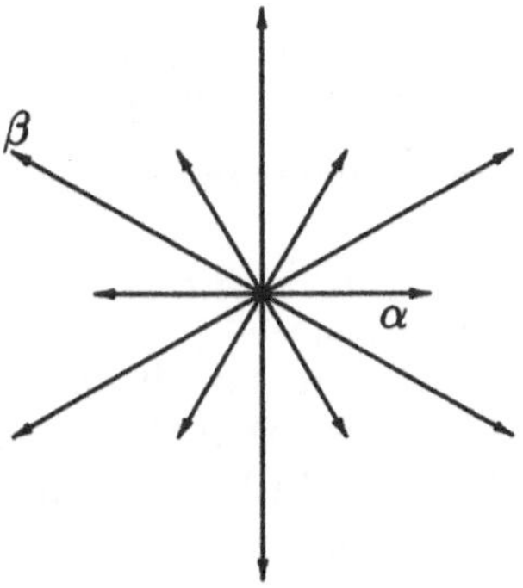

Ratio $|\beta| : |\alpha| = \sqrt{3}$; angle between adjacent vectors $= \pi/6$

Figure 2

(The usual metric in the plane is intended.)

Comment: The names for these figures are chosen, because these are the root systems of the corresponding Lie algebras in the Cartan-Killing classification (G_2 refers to an "exceptional" Lie algebra, see §2.14.)

Proof: Type $A_1 \oplus A_1$ clearly corresponds to the case of a decomposable (not simple) root system of rank 2. We turn to the simple case. One verifies easily that figures (ii), (iii), (iv) above are root systems, i.e., that conditions (i), (ii), (iii) of §2.6 are satisfied. The Weyl groups are the dihedral groups $\mathcal{D}_3, \mathcal{D}_4, \mathcal{D}_6$. In each case the reflections S_α and S_β, for the given α and β, generate the Weyl group; also, the whole system is generated by applying the Weyl group to the two vectors α and β. — We must show that there are no other systems:

Let a simple root system of rank two be given. Choose a shortest vector α, and let β be another vector, independent of, but not orthogonal to α (this must exist). Applying S_α, if necessary, we may assume $\langle \alpha, \beta \rangle < 0, i.e., a_{\beta\alpha} < 0$. We then have the possibilities in Fig.1 for the pair α, β. In cases (iii) and (iv) we know already that the reflections S wr to α and β will generate the systems B_2 and G_2; and it is clear that there can't be any other vectors in the system because of the restrictions on angles and norms from the table above. In case (ii) α and β generate A_2; and the only way to have more vectors in the system is to go to G_2, again because of the restrictions on angles and norms. $\surd$

The importance of the rank 2 case stems from the following simple observation: If R is a root system in the space V, and W is a subspace of V, then $R \cap W$, if not empty, is a root system in W. Thus, if α and β are any two independent vectors in R, the intersection of the plane $((\alpha, \beta))$ with R is one of our four types. (In case $A_1 \oplus A_1$, i.e., α orthogonal to β and $\alpha + \beta$ not in R, one calls α and β *strongly orthogonal.*)

A glance at figures (i) - (iv) shows

PROPOSITION B. *Let α and β be two elements of a root system R (with $\beta \neq 1\alpha$), and put $\varepsilon = \operatorname{sign} a_{\beta\alpha}$. Then all the elements β, $\beta - \varepsilon\alpha, \beta - 2\varepsilon\alpha, \ldots, \beta - a_{\beta\alpha} \cdot \alpha$ belong to R; in particular, if $\langle \alpha, \beta \rangle > 0$, then $\beta - \alpha$ belongs to R.*

Note: These α, β don't have to correspond to the α and β in the figures, but can be any two (independent) vectors. For the roots of a Lie algebra we met this in Prop. B and B″ of §2.5. Note that the axioms for root systems require only that the ends of the chain in Prop. B belong to R. The dots $\ldots$ in the chain are of course slightly misleading; it is clear from the figures that there are at most four terms in any chain. In fact, one reads off: The α-string of β (defined as in §2.4 as the set of elements of R of the form $\beta + t\alpha$ with integral t) is unbroken, i.e., t runs exactly through some interval $-q \leq t \leq p$ with p, q non-negative integers; and it contains at most four vectors.

2.8 Weyl-Chevalley normal form, first stage

We continue with a semisimple Lie algebra $\mathfrak{g}$, with $CSA\mathfrak{h}$, root system Δ, etc., as described in the preceding sections. Our aim is to show that Δ determines $\mathfrak{g}$. Roughly speaking this amounts to showing the existence of a basis for $\mathfrak{g}$, such that the corresponding structure constants can be read off from Δ; this is the *Weyl-Chevalley normal form* (Theorem A, §2.9). The present section brings a preliminary step.

For each root α choose a root element X_α in $\mathfrak{g}_\alpha$, subject to the condition $[X_\alpha X_{-\alpha}] = H_\alpha$ (see §2.5); these vectors, suitably normalized, will be part of the Weyl-Chevalley basis. For any two α, β in Δ with $\beta \neq \pm\alpha$ we have $[X_\alpha X_\beta] = N_{\alpha\beta} X_{\alpha+\beta}$, with some coefficient $N_{\alpha\beta}$ in $\mathbb{C}$, by $\mathfrak{g}_\alpha = ((X_\alpha))$ (Prop. A of §2.4) and (2) of §2.2. We also put $X_\lambda = 0$, if λ is an element of $\mathfrak{h}^\top$ not in Δ; and we put $N_{\lambda\mu} = 0$ for λ and μ in $\mathfrak{h}^\top$ and at least one of

$\lambda, \mu, \lambda+\mu$ not a root. Our aim is to get fairly explicit values for the $N_{\alpha\beta}$ by suitable choice of the X_α. The freedom we have is to change each X_α by a factor c_α (as long as we have $c_{-\alpha} = 1/c_\alpha$, to preserve $[X_\alpha X_{-\alpha}] = H_\alpha$).

Let α, β be two roots, with $\beta \neq \pm\alpha$. Let the α-string of β go from $\beta - q\alpha$ to $\beta + p\alpha$ (see §2.4). The main observation is the following proposition, which ties down the $N_{\alpha\beta}$ considerably.

PROPOSITION A. $N_{\alpha\beta} \cdot N_{-\alpha,-\beta} = -(q+1)^2$, *if* $\alpha + \beta$ *is a root.*

For the proof we first develop two formulae.

(1) $N_{\alpha\beta} = -N_{\beta\alpha}$ for any two roots α, β.

This is immediate from skew symmetry of [ ].

(2) $N_{\alpha\beta}/\langle\gamma,\gamma\rangle = N_{\beta\gamma}/\langle\alpha,\alpha\rangle = N_{\gamma\alpha}/\langle\beta,\beta\rangle$ for any three pairwise independent roots α, β, γ with $\alpha+\beta+\gamma = 0$.

Proof of (2): From the Jacobi identity $[X_\alpha[X_\beta X_\gamma]] + \ldots = 0$ we get $N_{\beta\gamma}H_\alpha + N_{\gamma\alpha}H_\beta + N_{\alpha\beta}H_\gamma = 0$ (note $[X_\beta X_\gamma] = N_{\beta\gamma}X_{-\alpha}$ and $[X_\alpha X_{-\alpha}] = H_\alpha$ etc.). On the other hand, the relation $\alpha+\beta+\gamma = 0$ implies the relation $h_\alpha + h_\beta + h_\gamma = 0$, and this in turn becomes $\langle\alpha,\alpha\rangle H_\alpha + \langle\beta,\beta\rangle H_\beta + \langle\gamma,\gamma\rangle H_\gamma = 0$. The coefficients of the two relations between the H's must be proportional, because of the pairwise independence of the H's.

We now prove Proposition A. Consider the representation ad of $\mathfrak{g}^{(\alpha)} = ((H_\alpha, X_\alpha, X_{-\alpha}))$ on the string $\mathfrak{g}^\alpha_\beta$. As noted after Prop. A in §2.5, this is equivalent to the representation D_s of A_1 with $2s = p+q$. One verifies that X_β corresponds to the vector v_p in the notation of §1.11. Recalling the formulae $X_+ v_i = \mu_i v_{i-1}$, with $\mu_i = i(2s+1-i)$, and $X_- v_{i-1} = v_i$, we get $\operatorname{ad} X_{-\alpha} \circ \operatorname{ad} X_\alpha(X_\beta) = \mu_p X_\beta = p(q+1)X_\beta$. The left-hand side of this equation transforms (with the help of (1) and (2) above) into $N_{\alpha\beta}[X_{-\alpha}X_{\alpha+\beta}] = N_{\alpha\beta}N_{\alpha,\alpha+\beta}X_\beta = -N_{\alpha\beta}N_{\alpha+\beta,-\alpha}X_\beta =$ $-N_{\alpha\beta}N_{-\alpha,-\beta} \cdot (\langle\beta,\beta\rangle/\langle\alpha+\beta,\alpha+\beta\rangle)X_\beta$.

Thus we have

$$N_{\alpha\beta}N_{-\alpha,-\beta} = -p(q+1)\langle\alpha+\beta,\alpha+\beta\rangle/\langle\beta,\beta\rangle$$

To get the value in Prop.A, we have to show $p\langle\alpha+\beta,\alpha+\beta\rangle = (q+1)\langle\beta,\beta\rangle$. As noted before, $((\alpha,\beta)) \cap \Delta$ is a rank two root system, necessarily simple in our case since $\alpha+\beta$ belongs to it. Thus we only have to go through the three root systems A_2, B_2, G_2 and to take for α and β any two vectors whose sum is also in the figure and check the result. We can of course work modulo the symmetry given by the Weyl group. We shall not go into the details. As an example take for α, β the vectors so named in G_2 in Prop.A of §2.7. We see $q = 0, p = 3$, and $\langle\beta,\beta\rangle = 3\langle\alpha+\beta,\alpha+\beta\rangle$ (see the table in §2.7 for the last equation).

We note an important consequence.

COROLLARY B. *If $\alpha + \beta$ is a root, then $N_{\alpha\beta}$ is not 0.*

2.9 Weyl-Chevalley normal form

The result we are getting to is a choice of the X_α for which the $N_{\alpha\beta}$ take quite explicit values. Historically this came about in steps, with Weyl [25, 26] and others proving first existence of real $N_{\alpha\beta}$'s and eventually narrowing this down to values in an extension of the rationals by square roots of rationals, and with Chevalley [6] taking the last big step, which made them explicit and showed them to be integers. We state the result as the *Weyl-Chevalley normal form*:

THEOREM A. *Let $\mathfrak{g}$ be a complex semisimple Lie algebra, with CSA $\mathfrak{h}$, root system Δ in $\mathfrak{h}_0^\top$, etc., as in the preceding sections.*

(i) There exist root elements X_α (generators of the $\mathfrak{g}_\alpha$), for all α in Δ, satisfying $[X_\alpha, X_{-\alpha}] = H_\alpha$, such that $[X_\alpha X_\beta] = \pm\ (q+1)X_{\alpha+\beta}$.

(ii) The $\pm$-signs in (i) are well-determined, up to multiplication by factors $u_\alpha u_\beta u_{\alpha+\beta}$, where the u_α are ± 1, arbitrary except for $u_{-\alpha} = u_\alpha$.

(iii) The X_α are determined up to factors c_α, arbitrary except for the conditions $c_\alpha \cdot c_{-\alpha} = 1$ and $c_\alpha \cdot c_\beta = \pm c_{\alpha+\beta}$.

Property (i), in detail, says that we have $N_{\alpha\beta} = \pm(q+1)$ for any two roots α, β with $\alpha + \beta$ also a root, with q the largest integer t such that $\beta - t\alpha$ is a root.

COROLLARY B. *There exists a basis for $\mathfrak{g}$, such that all structure constants are integers ($\mathfrak{g}$ has a $\mathbb{Z}$-form).*

COROLLARY C (THE ISOMORPHISM THEOREM). *Let $\mathfrak{g}_1$ and $\mathfrak{g}_2$ be two semisimple Lie algebras over $\mathbb{C}$, with root systems Δ_1 and Δ_2. If Δ_1 and Δ_2 are weakly equivalent, in the sense that there exists a bijection $\varphi : \Delta_1 \to \Delta_2$ that preserves the additive relations (i.e., $\varphi(-\alpha) = -\varphi(\alpha)$, and whenever α, β, and $\alpha+\beta$ belong to Δ_1, then $\varphi(\alpha+\beta) = \varphi(\alpha) + \varphi(\beta)$, and similarly for φ^{-1}), then $\mathfrak{g}_1$ and $\mathfrak{g}_2$ are isomorphic.*

We shall comment on the corollaries after the proof of the main result. We begin by noting that by Prop.A in §2.8 for any pair α, β in Δ with $\alpha + \beta$ also a root the relation $N_{\alpha\beta} = \pm(q+1)$ is equivalent to the relation

$$(*) \qquad N_{\alpha\beta} = -N_{-\alpha,-\beta}.$$

For the proof of Theorem A we shall show that one can adjust the original X_α so that $(*)$ holds for all α and β. This will be done inductively wr to a (weak) *order* in $\mathfrak{h}_0^\top$, defined as follows: Choose an element H_0 in $\mathfrak{h}_0$ with $\alpha(H_0) \neq 0$ for all roots α (this clearly exists) and for any λ, μ in $\mathfrak{h}_0^\top$ define $\lambda > \mu$ to mean $\lambda(H_0) > \mu(H_0)$, and also $\lambda \geq \mu$ to mean $\lambda(H_0) \geq \mu(H_0)$.

Clearly the relation $>$ is transitive (and irreflexive); but note that $\lambda \geq \mu$ and $\lambda \leq \mu$ together do not imply $\lambda = \mu$. We use obvious properties such as: If $\lambda > 0$, then $\lambda + \mu > \mu$. We describe $\lambda > 0$ as "λ is positive", etc. (One can and often does refine this weak order to a total order on $\mathfrak{h}_0$, defined by lexicographical order of the components wr to some basis.) We write Δ^+ for the set of positive roots, i.e.those roots that are > 0 in this order; similarly Δ^- is the set of negative roots. Clearly Δ^- is simply $-\Delta^+$, and Δ is the disjoint union of Δ^+ and Δ^-.

We first reduce the problem to the positive roots.

LEMMA D. *(i) If relation $(*)$ holds whenever α and β are positive, then it holds for all α and β;*

(ii) Let λ, in $\mathfrak{h}^\top$, be positive. If $(*)$ holds for all positive α and β with $\alpha + \beta < \lambda$, then it also holds for all negative α and β with $\alpha + \beta > -\lambda$ and for all α and β with $0 < \alpha < \lambda$ and $0 > \beta > -\lambda$.

We prove (ii); the proof for (i) results by omitting all references to λ. The case where α and β are both negative follows trivially. Let then α and β be given as in the second part of (ii), and put $\gamma = -\alpha - \beta$. Say $\gamma < 0$; then we have $0 > \gamma + \beta = -\alpha > -\lambda$. From the hypothesis and §2.8, (1) and (2), we find $N_{\alpha\beta}/\langle\gamma,\gamma\rangle = N_{\beta\gamma}/\langle\alpha,\alpha\rangle = -N_{-\beta,-\gamma}/\langle\alpha,\alpha\rangle = N_{-\gamma,-\beta}/\langle\alpha,\alpha\rangle = N_{-\beta,-\alpha}/\langle\gamma,\gamma\rangle = -N_{-\alpha,-\beta}/\langle\gamma,\gamma\rangle$, i.e., $(*)$ holds for α and β.

Note that $(*)$ holds trivially for $N_{\lambda\mu}$ with λ or μ or $\lambda + \mu$ not a root.

The induction step for the proof of (*) is contained in the next computation.

LEMMA E. *Let η be a positive root, and suppose that $(*)$ holds for all pairs of positive roots with sum $< \eta$. Let $\gamma, \delta, \varepsilon, \zeta$, be four positive roots with $\gamma + \delta = \varepsilon + \zeta = \eta$. Then the relation*

$$N_{\gamma\delta}/N_{-\gamma,-\delta} = N_{\varepsilon\zeta}/N_{-\varepsilon,-\zeta}$$

holds.

For the proof we may assume $\gamma \geq \varepsilon \geq \zeta \geq \delta$. We write out the Jacobi identity for $X_\gamma, X_\delta, X_{-\varepsilon}$: $0 = [X_\gamma[X_\delta X_{-\varepsilon}]] + \ldots = (N_{\delta,-\varepsilon}N_{\gamma,\zeta-\gamma}) + N_{-\varepsilon,\gamma}N_{\delta,\zeta-\delta} + N_{-\varepsilon,\varepsilon+\zeta}N_{\gamma\delta})X_\eta$.

Using §2.8, (1) and (2), we get the relation

$$(**) \qquad N_{\delta,-\varepsilon}N_{\gamma,\zeta-\gamma} + N_{-\varepsilon,\gamma}N_{\delta,\zeta-\delta} = N_{\gamma\delta}N_{-\varepsilon,-\zeta} \cdot \langle\zeta,\zeta\rangle/\langle\eta,\eta\rangle \ .$$

This relation also holds, of course, with all roots replaced by their negatives. Now under our induction hypothesis this replacement does not change the left-hand side of (**). Namely, first we have $N_{\delta,-\varepsilon} = -N_{-\delta,\varepsilon}$ and $N_{-\varepsilon,\gamma} = -N_{\varepsilon,-\gamma}$ by Lemma D (ii); secondly, if $\zeta-\gamma$ is a root at all, then it is clearly ≤ 0 and Lemma D (ii) applies again; similarly $N_{\delta,\zeta-\delta} = -N_{-\delta,\delta-\zeta}$. Therefore the right-hand side of (**) is also invariant under the change of sign of all the roots involved, and Lemma E follows.

Given roots η, γ, δ with $\eta = \gamma + \delta$ as in Lemma E (i.e., with (*) holding "below η"), we can multiply X_η by a suitable factor c_η (and $X_{-\eta}$ by $1/c_\eta$) so that (*) holds with γ, δ for α, β (so that $N_{\gamma\delta} = \pm(q+1)$; we can even prescribe the sign). It follows from Lemma E that then (*) holds automatically for all pairs ε, ζ with $\varepsilon + \zeta = \eta$. This is the induction step. (We induct over the finite set $\alpha(H_0)$, α in Δ^+, where H_0, in $\mathfrak{h}_0$, defines the order in $\mathfrak{h}_0$. The induction begins with the lowest positive roots; they are not sums of two positive roots.) This establishes part (i) of Theorem A.

Regarding the ambiguity of signs for the $N_{\alpha\beta}$ we note the following: suppose we choose for each positive root η a specific pair γ, δ with $\gamma+\delta = \eta$ (if such pairs exist) and we choose a sign for $N_{\gamma\delta}$ arbitrarily; then the signs of the other $N_{\varepsilon\zeta}$, for ε, ζ with $\varepsilon+\zeta = \eta$, are determined by (**) (inductively; note that the "mixed" N's, with one root positive and the other negative, in (**) are already determined, as in the proof of Lemma D, by (2) in §2.8). We refer to such a choice (of the γ, δ and the signs) as a *normalization*.

As for part (ii) of Theorem A, the statement about the X_α should be clear: since the $N_{\alpha\beta}$ are determined (up to sign), the freedom in the choice of the X_α amounts to factors c_α as indicated. For the signs of the $N_{\alpha\beta}$ it is clear that multiplying X_α by u_α results in multiplying $N_{\alpha\beta}$ by $u_\alpha u_\beta u_{\alpha+\beta}$. In the other direction, let $\{X'_\alpha, N'_{\alpha\beta}\}$ be another set of quantities as in Theorem A. Using a normalization, with the given N's, and arguing as in Lemmas D and E, one constructs the factors u_α inductively. At the "bottom" one can take them as 1; and (**) implies that adjusting $N'_{\gamma\delta}$ for the chosen pair γ, δ automatically yields agreement for the other ε, ζ with $\varepsilon + \zeta = \gamma + \delta$.√

We come to Corollary B. We choose as basis the X_α of the normal form, together with any l independent ones of the H_α. We then have $[H_\alpha H_\beta] = 0$, $[H_\alpha X_\beta] = a_{\beta\alpha} X_\beta$, $[X_\alpha X_\beta] = N_{\alpha\beta} X_{\alpha+\beta}$. √

Next the important Corollary C. Note that the map φ, the weak equivalence of Δ_1 and Δ_2, is not assumed to be a linear map, but only a map between the finite sets Δ_1 and Δ_2, preserving the relations of the two types $\alpha + \beta = 0$ and $\gamma = \alpha + \beta$. Now the Cartan integers are determined by these relations, through the notion of strings and formula (5) of §2.5; thus we have $a^1_{\alpha\beta} = a^2_{\varphi(\alpha)\varphi(\beta)}$ for all roots α and β. The Cartan integers in turn determine the inner products $\langle H_\alpha, H_\beta \rangle$, by $a_{\alpha\beta} = \alpha(H_\beta)$ and formula (8) of §2.5 for $\langle \cdot, \cdot \rangle$; these in turn determine the $\langle \alpha, \beta \rangle (= \langle h_\alpha, h_\beta \rangle)$

by $\langle H_\alpha, H_\alpha\rangle = 4/\langle h_\alpha, h_\alpha\rangle$ and $a_{\beta\alpha} = 2\langle h_\beta, h_\alpha\rangle/\langle h_\alpha, h_\alpha\rangle$. Thus the map φ from Δ_1 to Δ_2 is an isometry. It therefore extends to a (linear) isometry of $\mathfrak{h}_{20}^\top$ to $\mathfrak{h}_{10}^\top$ (the linear map that sends some l independent ones of the α's to their φ-images is an isometry, and thus sends every α to $\varphi(\alpha)$). This map extends to a $\mathbb{C}$-linear map of $\mathfrak{h}_2^\top$ to $\mathfrak{h}_1^\top$, whose transpose in turn is an isomorphism, again denoted by φ, of $\mathfrak{h}_1$ and $\mathfrak{h}_2$; it clearly preserves the Killing form and sends the coroots $H_{1,\alpha}$ to the coroots $H_{2,\varphi(\alpha)}$. We now take $\mathfrak{g}_1$ and $\mathfrak{g}_2$ in Weyl-Chevalley normal form. Then Theorem A implies that the $N_{1,\alpha\beta}$ equal the $N_{2,\varphi(\alpha)\varphi(\beta)}$, provided one is careful about the signs. That this is correct up to sign follows from the fact that the q-values entering into the N's are determined by the additive relations between the roots, and these are preserved by φ. To get the signs to agree, we choose the weak order and normalization for $\mathfrak{h}_2$ as the φ-images of those for $\mathfrak{h}_1$. Finally we define a linear map Φ from $\mathfrak{g}_1$ to $\mathfrak{g}_2$ by $\Phi|\mathfrak{h}_1 = \varphi$, and $\Phi(X_{1,\alpha}) = X_{2,\varphi(\alpha)}$. It is clear that this is a Lie algebra isomorphism, since it preserves the $a_{\alpha\beta}$ and the $N_{\alpha\beta}$ (see the formulae in Corollary B). $\checkmark$

To put the whole matter briefly: The normal form describes $\mathfrak{g}$ so explicitly in terms of the set of roots Δ (up to some ambiguity in the signs) that a weak equivalence of Δ_1 and Δ_2 induces (although not quite uniquely) an isomorphism of $\mathfrak{g}_1$ and $\mathfrak{g}_2$.

Examples for the isomorphism theorem, with $\mathfrak{g}_1 = \mathfrak{g}_2 = \mathfrak{g}$:

(a) The map $\alpha \to -\alpha$ is clearly a weak equivalence of Δ with itself; "the" corresponding automorphism of $\mathfrak{g}$ is $-id$ on $\mathfrak{h}$ and sends X_α to $-X_{-\alpha}$. One can work this out from the general theory, or, simpler, verify directly that this map is an automorphism. Note that it is an involution, i.e., its square is the identity map. It is related to the "normal real form" of $\mathfrak{g}$, see §2.10. We call it the (abstract) *contragredience* or *duality* and denote it by $C^\vee$; it is also called the *Chevalley involution.* For $A_n = \mathfrak{sl}(n+1, \mathbb{C})$, with a suitable $\mathfrak{h}$, it is the usual contragredience $X \to X^\triangle = -X^\top$.

(b) Take β in Δ, and let S_β be the corresponding Weyl reflection in $\mathcal{W}$. Since S_β is linear, it defines a weak equivalence of Δ with itself. The corresponding automorphism A_β will send X_α to $\pm X_{\alpha'}$, with $\alpha' = S_\beta(\alpha)$. There are likely to be some minus signs, since S_β will not preserve the weak order.

2.10 Compact form

A real Lie algebra is called *compact* if its Killing form is definite (automatically negative definite: invariance of κ implies that the $\operatorname{ad} X$ are skew-symmetric operators and have therefore purely imaginary eigenvalues; the eigenvalues and the trace of $\operatorname{ad} X \circ \operatorname{ad} X$ are then real and < 0).

A real Lie algebra $\mathfrak{g}_0$ is is called a *real form* of a complex Lie algebra $\mathfrak{g}$, if $\mathfrak{g}$ is (isomorphic to) the complexification of $\mathfrak{g}_0$.

Note that $\mathfrak{g}$ may have several non-isomorphic (over $\mathbb{R}$) real forms. Example: The real orthogonal Lie algebra $\mathfrak{o}(n) = \mathfrak{o}(n, \mathbb{R})$ is compact (verify that $\operatorname{ad} X$ is skew-symmetric on $\mathfrak{o}(n)$ wr to the usual inner product $\operatorname{tr} M^\top N$ on matrix space, or work out the Killing form). It is a real form of the orthogonal Lie algebra $\mathfrak{o}(n, \mathbb{C})$; every complex matrix M with $M^\top + M = 0$ is uniquely of the form $A + iB$ with A and B in $\mathfrak{o}(n)$, and conversely. Now let $I_{p,q}$ be the matrix $\operatorname{diag}(1, \ldots, 1, -1, \ldots, -1)$ with p 1's and $q(= n-p)$ -1's. Then the $\mathfrak{o}(p, q) = \{M : M^\top I_{p,q} + I_{p,q} M = 0\}$ are other real forms of $\mathfrak{o}(n, \mathbb{C})$.

$\mathfrak{o}(p, q)$ consists of the operators in $\mathbb{R}^n$ that leave the indefinite form $x_1^2 + \ldots + x_p^2 - x_{p+1}^2 - \ldots - x_n^2$ invariant (infinitesimally). Actually this is an "abstract" real form of $\mathfrak{o}(n, \mathbb{C})$, *i.e.*, $\mathfrak{o}(p, q) \otimes \mathbb{C}$ is *isomorphic* to $\mathfrak{o}(n, \mathbb{C})$, but $\mathfrak{o}(p, q)$ is not *contained* in $\mathfrak{o}(n, \mathbb{C})$ as real sub Lie algebra. To remedy this we should apply the coordinate transformation $x_r = x'_r$ for $r = 1, \ldots, p$ and $x_r = ix'_r$ for $r = p+1, \ldots, n$. This changes our quadratic form into the usual sum of squares, and transforms $\mathfrak{o}(p, q)$ into a real sub Lie algebra $\mathfrak{o}_0$ of $\mathfrak{o}(n, \mathbb{C})$, which is a real form in the concrete sense that $\mathfrak{o}(n, \mathbb{C})$ equals $\mathfrak{o}_0 + i\mathfrak{o}_0$ (over $\mathbb{R}$).)

As a matter of fact, the $\mathfrak{o}(p, q)$ together with one more case represent all possible real forms of $\mathfrak{o}(n, \mathbb{C})$. The additional case, $\mathfrak{o}^*(2n)$, exists only for even dimension and consists of all matrices in $\mathfrak{o}(2n, \mathbb{C})$ that satisfy $M^* J + JM = 0$ (where * means transpose conjugate = adjoint, and J is the matrix of §1.1).

We come to an important fact, discovered by H. Weyl (and in effect known to E. Cartan earlier, via the Killing-Cartan classification; we might note here the peculiar phenomenon that many facts about semisimple Lie algebras were first verified for all the individual Lie algebras on the list, with a general proof coming later).

THEOREM A. *Every complex semisimple Lie algebra has a compact real form .*

The proof is an explicit description of this form, starting from the Weyl-Chevalley normal form; we use the notation developed above. (For an alternate proof without the normal form see R. Richardson [21].)

Let $\mathfrak{u}$ be the real subspace of $\mathfrak{g}$ spanned by $i\mathfrak{h}_0$ and the elements $U_\alpha = X_\alpha - X_{-\alpha}$ and $V_\alpha = i(X_\alpha + X_{-\alpha})$ with α running over the positive roots (for the given choice of $>$). We see at once that $\dim_{\mathbb{R}} \mathfrak{u} \le \dim_{\mathbb{C}} \mathfrak{g}$, and that $\mathfrak{u}$ spans $\mathfrak{g}$ over $\mathbb{C}$ (we get all of $\mathfrak{h} = \mathfrak{h}_0 + i\mathfrak{h}_0$, and we can "solve" for the X_α and $X_{-\alpha}$); this shows that at any rate $\mathfrak{u}$ is a real form of $\mathfrak{g}$ as vector space.

That $\mathfrak{u}$ is a sub Lie algebra (and therefore a real form of $\mathfrak{g}$ as Lie algebra) is a simple verification. For example: $[iHU_\alpha] = \alpha(H) V_\alpha$ and $[iHV_\alpha] =$

$-\alpha(H)U_\alpha$ (note $\alpha(H)$ is real for H in $\mathfrak{h}_0$); for $[U_\alpha V_\beta]$ one has to make use of $N_{\alpha\beta} = -N_{-\alpha,-\beta}$.

Finally, the Killing form: We have $\langle X_\alpha, X_{-\alpha}\rangle = 2/\langle\alpha,\alpha\rangle$ (see end of §2.4 and recall $[X_\alpha X_{-\alpha}] = H_\alpha$). From this and from $\langle X_\alpha, X_\beta\rangle = 0$ unless $\beta = -\alpha$ (see (3) in §2.2) one computes: for $X = iH + \sum r_\alpha U_\alpha + \sum s_\alpha V_\alpha$ with H in $\mathfrak{h}_0$ and r_α, s_α real one has

$$\langle X, X\rangle = -\sum_{\Delta} \alpha(H)^2 - 4\sum_{\Delta+}(r_\alpha^2 + s_\alpha^2)/\langle\alpha,\alpha\rangle.$$

(The first sum over all roots, the second one over the positive ones.)

Clearly the form is negative definite, and so $\mathfrak{u}$ is a compact real form. We will see soon that up to automorphisms of $\mathfrak{g}$ there is only one compact real form.

(The importance of the compact form comes from the theorem of H.Weyl that any Lie group to this Lie algebra is automatically compact. This makes integration over the group very usable; it is the basis of Weyl's original, topological-analytical proof for complete reducibility of representations (§3.4).)

The next theorem shows how to construct all real forms of $\mathfrak{g}$ from facts about the compact real form. The main ingredient are involutory automorphisms of $\mathfrak{u}$.

THEOREM B. *Let $\mathfrak{u}$ be a compact form of $\mathfrak{g}$. (i) Given an involutory automorphism A of $\mathfrak{u}$, let $\mathfrak{k}$ and $\mathfrak{p}$ be the $+1-$ and $-1-$ eigenspaces of A. Then the real subspace $\mathfrak{k} + i\mathfrak{p}$ of $\mathfrak{g}$ is a real form of $\mathfrak{g}$. (ii) Every real form of $\mathfrak{g}$ is obtained this way, up to an automorphism of $\mathfrak{g}$ (which can be taken of the form* $\exp(\operatorname{ad} X_0)$ *with some X_0 in $\mathfrak{g}$).*

Thus, in order to find the real forms of $\mathfrak{g}$, one should find the involutions of $\mathfrak{u}$ – usually a fairly easy task.

Proof: Let A be an involution of $\mathfrak{u}$. The equation $A^2 = id$ implies by standard arguments that the eigenvalues of A are $+1$ and -1, and that $\mathfrak{u}$ is direct sum of the corresponding eigenspaces $\mathfrak{k}$ and $\mathfrak{p}$. From $A[XY] = [AX, AY]$ one reads off the important relations

(1) $$[\mathfrak{k}\mathfrak{k}] \subset k, [\mathfrak{k}\mathfrak{p}] \subset \mathfrak{p}, [\mathfrak{p}\mathfrak{p}] \subset \mathfrak{k}.$$

In particular, $\mathfrak{k}$ is a sub Lie algebra.

Now $\mathfrak{k} + i\mathfrak{p}$ is a real form of $\mathfrak{g}$ as vector space, since it spans $\mathfrak{g}$ over $\mathbb{C}$ just as much as $\mathfrak{u}$ does and its $\mathbb{R}-$ dimension equals that of $\mathfrak{u}$ (note $\mathfrak{u} \cap i\mathfrak{u} = 0$). From (1) one concludes that $\mathfrak{k} + i\mathfrak{p}$ is a (real) subalgebra: besides $[\mathfrak{k}\mathfrak{k}] \subset \mathfrak{k}$ we have $[\mathfrak{k}, i\mathfrak{p}] = i[\mathfrak{k}\mathfrak{p}] \subset i\mathfrak{p}$ and $[i\mathfrak{p}, i\mathfrak{p}] = -[\mathfrak{p}\mathfrak{p}] \subset \mathfrak{k}$. (It is important that [ ] is $\mathbb{C}$-linear.) This establishes part (i) of Theorem A.

We note that the step from the involution A to the direct sum decomposition $\mathfrak{u} = \mathfrak{k} + \mathfrak{p}$ with relations (1) holding is reversible: If one has such a decomposition of $\mathfrak{u}$, one defines A by $A|\mathfrak{k} = id$ and $A|\mathfrak{p} = -id$. This is clearly an involutory linear map, and (1) implies immediately that it preserves brackets. Then A preserves the Killing form, and it follows that $\mathfrak{k}$ and $\mathfrak{p}$ are orthogonal to each other (via $\langle X, Y\rangle = \langle AX, AY\rangle = \langle X, -Y\rangle = -\langle X, Y\rangle$ for X in $\mathfrak{k}$ and Y in $\mathfrak{p}$).

The proof of part (ii) is more complicated. First we introduce the notion of (complex) *conjugation.* (Cf.§1.4.) Let V_0 be a real form of the (complex) vector space V (so that every vector X of V is uniquely of the form $X'+iX''$ with X', X'' in V_0). Then the conjugation of V wr to V_0 is the conjugate-linear map σ of V to itself given by $\sigma(X'+iX'') = X' - iX''$. ("Conjugate-linear" means $\sigma(a \cdot v) = \bar{a} \cdot v$ for a in $\mathbb{C}$, v in V.) Note that σ is of order two, i.e., $\sigma^2 = id$, or $\sigma = \sigma^{-1}$.

Let now $\mathfrak{g}_0$ be a real form of our Lie algebra $\mathfrak{g}$. Let σ and τ be the conjugations of (the vector space) $\mathfrak{g}$ wr to its real forms $\mathfrak{g}_0$ and $\mathfrak{u}$ respectively. Both σ and τ are $\mathbb{R}$-automorphisms of $\mathfrak{g}$ (they are $\mathbb{R}$-linear and preserve brackets, as immediately verified using $X = X' + iX''$ etc.). The two compositions $\sigma \circ \tau$ and $\tau \circ \sigma$ are again $\mathbb{C}$-automorphisms.

The following observation is crucial: If σ and τ commute, then $\mathfrak{u}$ is σ-invariant, and conversely.

Indeed, if $\sigma \circ \tau = \tau \circ \sigma$, then σ preserves the +1-eigenspace of τ , which is precisely $\mathfrak{u}$. Conversely, if $\sigma(\mathfrak{u}) = \mathfrak{u}$, then also $\sigma(i\mathfrak{u}) = i\mathfrak{u}$, since σ is conjugate linear. Now $\tau|\mathfrak{u} = id$ and $\tau|i\mathfrak{u} = -id$, and so clearly σ and τ commute on $\mathfrak{u}$ and on $i\mathfrak{u}$, and so on $\mathfrak{g}$.

Our plan is now to replace $\mathfrak{g}_0$, via an automorphism of $\mathfrak{g}$, by another (isomorphic) real form $\mathfrak{g}_1$, whose associated conjugation σ_1 commutes with τ. And then the composition $\sigma_1 \circ \tau$ will function as the involution A of Theorem B.

The definition of real form implies that the Killing form κ of $\mathfrak{g}$ is simply the extension to complex coefficients of the Killing form of either $\mathfrak{g}_0$ or $\mathfrak{u}$; in particular κ is real on $\mathfrak{g}_0$ and on $\mathfrak{u}$. One concludes

$$(2) \qquad \kappa(\sigma X, \sigma Y) = \kappa(\tau X, \tau Y) = \kappa(X, Y)^- \text{ for all } X, Y \text{ in } \mathfrak{g}\,,$$

by writing $X = X' + iX'', Y = Y' + iY''$, and expanding. We introduce the sesquilinear form $\pi(X, Y) = \kappa(\tau X, Y)$ on $\mathfrak{g}$ (it is linear in Y and conjugate linear in X) and prove that it is negative definite Hermitean:

First, by (2) we have $\pi(Y, X) = \kappa(\tau Y, X) = \kappa(X, \tau Y) = \kappa(X, \tau^2 Y)^- = k(\tau X, Y)^- = \pi(X, Y)^-$. Second, writing again X as $X' + iX''$ with X', X'' in $\mathfrak{u}$, we have $\pi(X, X) = \kappa(X' - iX'', X' + iX'') = \kappa(X', X') + \kappa(X'', X'')$ (recall that κ is $\mathbb{C}$-bilinear); and κ is negative definite on $\mathfrak{u}$.

The automorphism $P = \sigma \circ \tau$ of $\mathfrak{g}$ is selfadjoint wr to π, by $\pi(PX, Y) = \kappa(\tau\sigma\tau X, Y) = \kappa(\sigma\tau X, Y)^- = \kappa(\tau X, \sigma\tau Y) = \pi(X, PY)$, using (2) twice.

Therefore the eigenvalues λ_i of P are real (and non-zero), and $\mathfrak{g}$ is the direct sum of the corresponding eigenspaces V_{λ_i}. From $P[XY] = [PX, PY]$ one concludes

$$(3) \qquad [V_{\lambda_i}, V_{\lambda_j}] \subset V_{\lambda_i \cdot \lambda_j} \qquad \text{(or } = 0\text{, if } \lambda_i \cdot \lambda_j \text{ is not eigenvalue of } P\text{).}$$

We introduce the operator $Q = |P|^{-1/2}$; that is, Q is multiplication by $|\lambda_i|^{-1/2}$ on V_{λ_i}. P and Q commute, of course. From (3) it follows that Q is a $\mathbb{C}$-automorphism of $\mathfrak{g}$. From $\lambda/|\lambda| = |\lambda|/\lambda$ for real $\lambda \neq 0$ we get $P \cdot Q^2 = P^{-1} \cdot Q^{-2}$.

We are ready to construct the promised real form $\mathfrak{g}_1$ of $\mathfrak{g}$, $\mathbb{R}$-isomorphic and conjugate to (i.e., image under an automorphism of $\mathfrak{g}$) $\mathfrak{g}_0$: We put $\mathfrak{g}_1 = Q(\mathfrak{g}_0)$. The conjugation σ_1 of $\mathfrak{g}$ wr to $\mathfrak{g}_1$ is clearly $Q{\cdot}\sigma{\cdot}Q^{-1}$. We want to prove that σ_1 and τ commute. We have $\sigma{\cdot}P{\cdot}\sigma^{-1} = \sigma{\cdot}\sigma{\cdot}\tau{\cdot}\sigma^{-1} = \tau{\cdot}\sigma = P^{-1}$, so that σ maps V_{λ_i} into V_{1/λ_i}. This implies $\sigma \cdot Q^{-1} \cdot \sigma^{-1} = Q$ (check the action on each V_{λ_i}). Then we have $\sigma_1 \cdot \tau = Q \cdot \sigma \cdot Q^{-1} \cdot \tau = Q^2 \cdot \sigma \cdot \tau = Q^2 \cdot P = P^{-1} \cdot Q^{-2} = \tau \cdot \sigma Q^{-2} = \tau \cdot Q \cdot \sigma \cdot Q^{-1} = \tau \cdot \sigma_1$; i.e., σ_1 and τ commute. $\checkmark$

As indicated above, this means that $\mathfrak{u}$ is stable under σ_1, and so the involutory automorphism $\sigma_1 \cdot \tau$ of $\mathfrak{g}$ restricts to an (involutory) automorphism, that we call A, of $\mathfrak{u}$. We split $\mathfrak{u}$ into the $+1-$ and $-1-$ eigenspaces of A, as $\mathfrak{k} + \mathfrak{p}$. Since $A = \sigma_1$ on $\mathfrak{u}$ and since $\mathfrak{g}_0$ is the $+1-$ eigenspace of σ_1 on $\mathfrak{g}$, it follows that we have $\mathfrak{k} \subset \mathfrak{g}_0$ (in fact,$\mathfrak{k} = \mathfrak{u} \cap \mathfrak{g}_0$). Since $\mathfrak{p}$ lies in the $-1-$ space of σ_1, the space $i\mathfrak{p}$ lies in the $+1-$ space of σ_1; so it too is contained in $\mathfrak{g}_0$. The sum $\mathfrak{k} + i\mathfrak{p}$ is direct (since $\mathfrak{k}$ and $\mathfrak{p}$ are $\mathbb{C}$-independent). For dimension reasons it must then equal $\mathfrak{g}_0$; this establishes Theorem B, with Q as the automorphism of $\mathfrak{g}$ involved, except for showing that Q is of the form $\exp(\operatorname{ad} X_0)$, an "inner" automorphism. To this end we note that the operator powers $|P|^t$ are defined for any real t (they are multiplication by $|\lambda_i|^t$ on V_{λ_i}; for $t = -1/2$ we get Q); they form a one-parameter subgroup, and are thus of the form $\exp(tD)$ with some derivation D of $\mathfrak{g}$, which by Cor.D, §1.10, is of the form $\operatorname{ad} X_0$ with some X_0 in $\mathfrak{g}$.

There are several important additions to this.

Corollary C. *Any two compact forms of $\mathfrak{g}$ are $\mathbb{R}$-isomorphic and conjugate in $\mathfrak{g}$.*

For the proof we note that the Killing form κ is positive definite on $i\mathfrak{p}$ (since it is negative definite on $\mathfrak{p}$). Therefore $\mathfrak{g}_0$, and $\mathfrak{g}_1$, are compact iff $\mathfrak{p} = 0$, that is iff $\mathfrak{g}_1 = \mathfrak{u}$ and $\mathfrak{u} = Q(\mathfrak{g}_0)$. — One speaks therefore of "the" compact form.

For a real form $\mathfrak{g}_0$ of $\mathfrak{g}$ a decomposition $\mathfrak{g}_0 = \mathfrak{k} + \mathfrak{p}$, satisfying the relations $[\mathfrak{k}\mathfrak{k}] \subset \mathfrak{k}, [\mathfrak{k}\mathfrak{p}] \subset \mathfrak{p}, [\mathfrak{p}\mathfrak{p}] \subset \mathfrak{k}$ and with the Killing form negative definite on $\mathfrak{k}$ and positive definite on $\mathfrak{p}$, is called a *Cartan decomposition* of $\mathfrak{g}_0$ (note that we have absorbed the earlier factor i into $\mathfrak{p}$). We restate part of Theorem B as follows.

THEOREM B′. *Every real form of* $\mathfrak{g}$ *has a Cartan decomposition.*

There is also a uniqueness statement. Suppose $\mathfrak{k}_1 + \mathfrak{p}_1$ and $\mathfrak{k}_2 + \mathfrak{p}_2$ are two Cartan-decompositions of the real form $\mathfrak{g}_0$, corresponding to the two compact forms $\mathfrak{u}_1 = \mathfrak{k}_1 + i\mathfrak{p}_1$ and $\mathfrak{u}_2 = \mathfrak{k}_2 + i\mathfrak{p}_2$.

PROPOSITION D. *There exists an automorphism* R *of* $\mathfrak{g}$*, of the form* $\exp(\operatorname{ad} X_0)$ *with some* X_0 *in* $\mathfrak{g}_0$*, that sends* $\mathfrak{k}_1$ *to* $\mathfrak{k}_2$ *and* $\mathfrak{p}_1$ *to* $\mathfrak{p}_2$*.*

Proof: Let σ, τ_1, τ_2 be the associated conjugations. As noted in the proof of Cor.C, the automorphism $R = |\tau_1 \cdot \tau_2|^{-1/2}$ sends $\mathfrak{u}_2$ to $\mathfrak{u}_1$. Now σ commutes with τ_1 and τ_2, and so with R, and so R maps $\mathfrak{g}_0$ to itself. We have $R(\mathfrak{k}_2) = R(\mathfrak{g}_0 \cap \mathfrak{u}_2) = R(\mathfrak{g}_0) \cap R(\mathfrak{u}_2) = \mathfrak{g}_0 \cap \mathfrak{u}_1 = \mathfrak{k}_1$, and similarly $R(\mathfrak{p}_2) = \mathfrak{p}_1$. The statement about the form of R follows similarly to the corresponding statement in Theorem B (ii), by considering the powers $|\tau_1 \cdot \tau_2|^t$. $\surd$

Clearly two involutions of $\mathfrak{u}$ that are conjugate in the automorphism group of $\mathfrak{u}$ give rise to two R-isomorphic real forms of $\mathfrak{g}$. The converse "uniqueness" fact also holds. Let A_1, A_2 be two involutions of $\mathfrak{u}$, with decompositions $\mathfrak{u} = \mathfrak{k}_1 + \mathfrak{p}_1 = \mathfrak{k}_2 + \mathfrak{p}_2$, and suppose the real forms $\mathfrak{g}_0 = \mathfrak{k}_1 + i\mathfrak{p}_1$, $\mathfrak{g}_2 = \mathfrak{k}_2 + i\mathfrak{p}_2$ are $\mathbb{R}$-isomorphic.

PROPOSITION E. *There exists an automorphism* B *of* $\mathfrak{g}$ *that sends* $\mathfrak{k}_1$ *to* $\mathfrak{k}_2$ *and* $\mathfrak{p}_1$ *to* $\mathfrak{p}_2$ *(and so* $BA_1B^{-1} = A_2$*).*

Proof: Let E be an isomorphism of $\mathfrak{g}_1$ with $\mathfrak{g}_2$. Then $E(\mathfrak{k}_1) + iE(\mathfrak{p}_1)$ is a Cartan decomposition of $\mathfrak{g}_2$, with associated compact form $E(\mathfrak{k}_1) + E(\mathfrak{p}_1)$. By Corollary C there is an automorphism Q of $\mathfrak{g}$ that sends $E(\mathfrak{k}_1)$ to $\mathfrak{k}_2$ and $E(\mathfrak{p}_1)$ to $\mathfrak{p}_2$. We can now take $Q \cdot E$ as B (regarding E as automorphism of $\mathfrak{g}$ by complexification). $\surd$

Altogether we have a bijection between the involutions of $\mathfrak{u}$ (up to automorphisms of $\mathfrak{u}$) and real forms of $\mathfrak{g}$ (up to isomorphism, or even conjugacy in $\mathfrak{g}$).

To look at a simple example, we let $\mathfrak{su}(n)$ be the $n \times n$ special-unitary (skew-Hermitean, trace 0) matrix Lie algebra (see §1.1). By explicit computation one finds that the Killing form is negative definite, so we have a semisimple compact Lie algebra. Let σ be the automorphism "complex conjugation"; it is involutory. The $+1-$ eigenspace consists of the real skew-symmetric matrices (this is the real orthogonal Lie algebra $\mathfrak{o}(n)$). Denoting the space of real symmetric matrices of trace 0 temporarily by $\mathfrak{s}(n)$, we can write the $-1-$ eigenspace of σ as $i\mathfrak{s}(n)$. Thus we have the decomposition $\mathfrak{su}(n) = \mathfrak{o}(n) + i\mathfrak{s}(n)$. The corresponding real form, obtained by multiplying the $\mathfrak{p}-$ part by i, is $\mathfrak{o}(n)+\mathfrak{s}(n)$. This is precisely the Lie algebra $\mathfrak{sl}(n,\mathbb{R})$ of all real $n \times n$ matrices of trace 0 (it being well known that any real matrix is uniquely the sum of a symmetric and a skew-symmetric one).

On the other hand, any complex matrix is uniquely of the form $A + iB$ with A and B Hermitean; we have therefore the direct sum decomposition $\mathfrak{sl}(n,\mathbb{C}) = \mathfrak{su}(n) + i\mathfrak{su}(n)$. Thus, finally, we can say that $\mathfrak{su}(n)$ is "the" compact real form of $\mathfrak{sl}(n,\mathbb{C})$ and that $\mathfrak{sl}(n,\mathbb{R})$ is a real form (there are still other real forms).

A real Lie algebra, equipped with an involutory automorphism, is called a *symmetric Lie algebra*; it is called *orthogonal symmetric*, if in addition it carries a definite quadratic form that is invariant (infinitesimally) under the ad X and under the involution. This is the infinitesimal version of E. Cartan's symmetric spaces and in particular Riemannian symmetric spaces. (See, e.g., [11,12,18].)

As an application of existence and uniqueness of compact real forms we prove

THEOREM F. *Any two Cartan sub algebras of a complex semisimple Lie algebra $\mathfrak{g}$ are conjugate in $\mathfrak{g}$ (under some inner automorphism of $\mathfrak{g}$).*

There exist more algebraic proofs, see [13, 24]. It is also possible to classify the Cartan sub algebras of real semisimple Lie algebras.

(We give G. Hunt's proof.) Let $\mathfrak{h}_1$ and $\mathfrak{h}_2$ be two CSAs of $\mathfrak{g}$. Each determines a compact form $\mathfrak{u}_i$ of $\mathfrak{g}$ as in Theorem A. By Corollary C we may assume $\mathfrak{u}_1 = \mathfrak{u}_2, = \mathfrak{u}$ say (replacing $\mathfrak{h}_2$ by a conjugate CSA). One verifies from the formulae after Theorem A that $i\mathfrak{h}_{1,0}$ and $i\mathfrak{h}_{2,0}$ are maximal Abelian sub Lie algebras of $\mathfrak{u}$. In fact, let H be an element of $\mathfrak{h}_{1,0}$ such that no root wr to $\mathfrak{h}_1$ vanishes on H (one calls such elements *regular* or *general*); then the centralizer of iH in $\mathfrak{u}$ is exactly $i\mathfrak{h}_{1,0}$.

The Killing form κ of $\mathfrak{g}$ is negative definite on $\mathfrak{u}$. Therefore the group G of all those operators on (the vector space) $\mathfrak{u}$ that leave κ invariant (a closed subgroup of $GL(\mathfrak{u})$) is compact; it is just the orthogonal group $O(\mathfrak{u},\kappa)$.

For X in $\mathfrak{u}$ the operators $\exp(t\cdot \operatorname{ad} X)$ are in G, by infinitesimal invariance of κ and the computation of §1.3. Let G_1 be the smallest closed subgroup of G that contains all the $\exp(\operatorname{ad} X)$; it is compact and all its elements are automorphisms of $\mathfrak{u}$ (and $\mathfrak{g}$). Now take general elements H_1 and H_2 of $\mathfrak{h}_{1,0}$ and $\mathfrak{h}_{2,0}$. On the orbit of iH_1 under G_1 (i.e., on the set $\{g(iH_1) : g \in G_1\}$) there exists by compactness a point with minimal distance (in the sense of κ) from iH_2. Since all the $\mathfrak{g}$ in G_1 are automorphisms of $\mathfrak{u}$, we may assume that iH_1 itself is that point (this amounts to replacing $\mathfrak{h}_1$ by its transform under some g in G_1). For any X in $\mathfrak{u}$ the curve $t \to \exp(t\cdot \operatorname{ad} X)(iH_1), = Y_t$ say, is then closest to iH_2 for $t = 0$. From $|Y_t - iH_2|^2 =$

$|Y_t|^2 - 2\langle Y_t, iH_2\rangle + |iH_2|^2$ and $|Y_t| = |iH_1|$ one sees that the derivative of $\langle \exp(t{\cdot}\mathrm{ad}\,X(iH_1)), iH_2\rangle$ vanishes for $t = 0$, so that we have $\langle [XH_1], H_2\rangle = 0$ for all X in $\mathfrak{u}$ (and then even in $\mathfrak{g}$). From $\langle [XH_1], H_2\rangle = \langle X, [H_1, H_2]\rangle$ and non-degeneracy of $\langle \cdot,\cdot\rangle$ we get $[H_1H_2] = 0$. This implies by the centralizer property above that iH_2 is contained in $i\mathfrak{h}_{1,0}$ and likewise iH_1 in $i\mathfrak{h}_{2,0}$, and then $i\mathfrak{h}_{1,0} = i\mathfrak{h}_{2,0}$ and also $\mathfrak{h}_1 = \mathfrak{h}_2$. $\surd$

We still have to show that the element $\mathfrak{g}$ used above is an *inner automorphism*, i.e., a finite product of $\exp(\mathrm{ad}\,X)$'s. This needs some basic facts about Lie groups that we shall not prove here: Let $A, \subset O(\mathfrak{u},\kappa)$, be the group of all automorphisms of $\mathfrak{u}$, and let A_0 be the *id*-component of A, a closed subgroup of course. Then A_0 is a Lie group; its Lie algebra (=tangent space at *id*) consists of the derivations of $\mathfrak{u}$, which by §1.10, Cor.D are all inner. This implies that A_0 is generated by the $\exp(\mathrm{ad}\,X)$ (with X in $\mathfrak{u}$) in the algebraic sense (i.e. the set of finite products is not only dense in A_0, but equal to it). Thus the group G_1 used above is identical with A_0, and the element $\mathfrak{g}$ is an inner automorphism. $\surd$

The argument proves in fact that all maximal Abelian sub Lie algebras of $\mathfrak{u}$ are conjugate in $\mathfrak{u}$, and that these sub Lie algebras are precisely the CSA's of $\mathfrak{u}$, i.e., the sub Lie algebras of $\mathfrak{u}$ that under complexification produce CSA's of $\mathfrak{g}$.

The definition of the rank of $\mathfrak{g}$ is now justified, since all CSA's clearly have the same dimension.

Another real form that occurs for every semisimple $\mathfrak{g}$ is the *normal real form*; it is defined by the requirement that for some maximal Abelian sub Lie algebra all operators $\mathrm{ad}\,X$ are real-diagonizable. In the Weyl-Chevalley normal form it is simply given by $\mathfrak{h}_0 + \sum_\Delta \mathbb{R}X_\alpha$. (For $\mathfrak{sl}(n,\mathbb{C})$ this turns out to be $\mathfrak{sl}(n,\mathbb{R})$.)

For any real form one defines the *character* as the signature of the Killing form (number of positive squares minus of negative squares in the diagonalized form). One can show that the character always lies between l (= rank of $\mathfrak{g}$) and $-n$ ($= -\dim_{\mathbb{C}}\mathfrak{g}$).

The compact form is the only real form with character $= -n$ and the normal real form the only one with character $= l$. (That these are the right values for the compact and normal real forms can be read off from the Weyl-Chevalley form of $\mathfrak{g}$.) We describe the arguments briefly. Given a Cartan decomposition $\mathfrak{k}+\mathfrak{p}$ of a real form $\mathfrak{g}_0$ (with the corresponding compact form $\mathfrak{u} = \mathfrak{k}+i\mathfrak{p}$ and the involution A of $\mathfrak{u}$ or $\mathfrak{g}$) one finds a maximal subspace $\mathfrak{a}$ of pairwise commuting elements of $\mathfrak{p}$ (by $[\mathfrak{pp}] \subset \mathfrak{k}$ this is the same as a maximal sub Lie algebra of $\mathfrak{p}$). One extends it to a maximal Abelian sub Lie algebra $\mathfrak{h}_0$ (= CSA) of $\mathfrak{u}$; it is of the form $\mathfrak{t} + i\mathfrak{a}$ with $\mathfrak{t}$ an Abelian sub Lie algebra of $\mathfrak{k}$. One also introduces the centralizer $\mathfrak{m} = \{X \in \mathfrak{k} : [X\mathfrak{a}] = 0\}$ of $\mathfrak{a}$ in $\mathfrak{k}$. Using the roots of $\mathfrak{g}$ wr to (the complexification of) $\mathfrak{h}_0$, one finds that $\mathfrak{a}$ has *general* elements, i.e., elements Y such that for any X in $\mathfrak{g}$ the relation

$[XY] = 0$ implies $[X\mathfrak{a}] = 0$. (The fundamental relations (1) for $\mathfrak{k}$ and $\mathfrak{p}$ show that the $\mathfrak{k}-$ and $\mathfrak{p}-$ components of such an X commute separately with Y.) The relation $\langle \operatorname{ad} Y.U, V\rangle + \langle U, \operatorname{ad} Y.V\rangle = 0$ shows that the linear transformations $\operatorname{ad} Y|\mathfrak{k}$ and $-\operatorname{ad} Y|\mathfrak{p}$ are adjoint wr to $\langle\cdot,\cdot\rangle$, and therefore have the same rank.

Thus we have $\dim \mathfrak{p} - \dim \mathfrak{a} = \dim \mathfrak{k} - \dim \mathfrak{m}$. It follows that the character of $\mathfrak{g}_0, = \dim \mathfrak{p} - \dim \mathfrak{k}$, equals $\dim \mathfrak{a} - \dim \mathfrak{m}$. Therefore it is at most equal to $\dim \mathfrak{h}_0, = l$; that it is at least equal to $-n$ is clear anyway. For the extreme case character $= l$ we have to have $\mathfrak{a} = \mathfrak{h}_0$ and $\mathfrak{m} = 0$ (i.e., $\mathfrak{p}$ contains a CSA of $\mathfrak{g}_0$). One can thus assume that the present $\mathfrak{h}_0$ is the sub Lie algebra of the Weyl-Chevalley normal form that there is called $i\mathfrak{h}_0$, and that the present $\mathfrak{u}$ is the $\mathfrak{u}$ there. Since $A|\mathfrak{h}$ is $-id$, we have $AX_\alpha = c_\alpha X_{-\alpha}$ (with $c_{-\alpha} = 1/c_\alpha$ of course); in the plane spanned by U_α and V_α (over $\mathbb{R}$) A induces a reflection. Conjugating A with a suitable inner automorphism one can arrange all the c_α to equal -1; then A is the "contragredience" of §2.9, and $\mathfrak{g}_0$ is the real form $\mathfrak{h}_0 + \sum \mathbb{R}X_\alpha$. $\surd$

The other extreme, character $= -n$, is simpler. We must have $\mathfrak{a} = 0$ (and $\mathfrak{m} = \mathfrak{u}$). But then $\mathfrak{p}$ is 0 (any non-zero X in $\mathfrak{p}$ spans a commutative sub Lie algebra), and so $\mathfrak{g}_0 = \mathfrak{u}$. $\surd$

2.11 Properties of root systems

We come to part C of our program (see §2.6).

Let R be a root system $\{\alpha, \beta, \ldots\}$ (see §2.6), in the (real) vector space V (with inner product $\langle\cdot,\cdot\rangle$); for simplicity assume $((R)) = V$. (Thus V corresponds to $\mathfrak{h}_0^\top$.) As in the case of the root system of $\mathfrak{g}$ in §2.9, we introduce a weak order $\geq$ in V by choosing an element v_0 of the dual space $V^\top$ that doesn't vanish at any α in R, and for any two vectors λ, μ defining $\lambda > \mu$ (resp. $\lambda \geq \mu$) to mean $v_0(\lambda) >$ (resp $\geq$) $v_0(\mu)$. This divides R into the two subsets R^+ and R^- of positive and negative elements. We define a root α, i.e., a vector α in R to be *simple* or *fundamental* if it is positive, but not sum of two positive vectors. (Note that this definition and all the following developments depend on the chosen ordering.)

Let $F = \{\alpha_1, \alpha_2, \ldots, \alpha_l\}$ be the set of all simple vectors in R; this is called the *simple* or *fundamental system* or also *basis* of R (wr to the given order in V). (In the case of the root system Δ of $\mathfrak{g}$ wr to $\mathfrak{h}$ we use Ψ to designate a fundamental system.) We derive some elementary, but basic properties of F.

PROPOSITION A.

(a) For distinct α and β in F one has $\langle\alpha, \beta\rangle \leq 0$;

(b) F is a linearly independent set;

(c) every positive element of R is linear combination of the fundamental vectors with non-negative integral coefficients;

(d) every non-simple vector in R^+ can be written as sum of two positive vectors of which at least one is simple.

Proof:

(a) If $\langle \alpha, \beta \rangle$ is positive, then $\alpha - \beta$ and $\beta - \alpha$ belong to R (see §2.7); say $\alpha - \beta$ belongs to R^+. Then $\alpha = \beta + (\alpha - \beta)$ contradicts simplicity of α.

(b) A relation $\sum x_i \alpha_i = 0$ can be separated into $\sum y_i \alpha_i = \sum z_j \alpha_j$ with all coefficients non-negative. Calling the left side λ and the right side μ, we get $0 \leq \langle \lambda, \lambda \rangle = \langle \lambda, \mu \rangle \leq 0$ (the last step by (a) upon expanding); thus $\lambda = \mu = 0$. But then $v_0(\lambda) = v_0(\mu) = 0$ implies that all y_i and z_i vanish.

(c) If α in R^+ is not simple, it is, by definition, sum of two vectors in R^+. If either of these is not simple, it in turn splits into two positive vectors. This can be iterated. Since the v_0-values clearly go down all the time, eventually all the terms must be simple.
This shows that F spans V and that l $(= \#F)$ equals $\dim V$. It is also fairly clear from (b) and (c) that F can be characterized as a linearly independent subset of R^+ such that R^+ lies in the cone spanned by it.

(d) By (c) there is an equation $\alpha = \sum n_i \alpha_i$ with non-negative integral coefficients. From $0 < \langle \alpha, \alpha \rangle = \sum n_i \langle \alpha, \alpha_i \rangle$ it follows that some $\langle \alpha, \alpha_i \rangle$ must be positive. Then $\alpha - \alpha_i$ belongs to R, by Proposition B of §2.7, and so either $\alpha - \alpha_i$ or $\alpha_i - \alpha$ is in R^+. But the latter can't be in R^+, since $\alpha_i = \alpha + (\alpha_i - \alpha)$ contradicts simplicity of α_i. $\surd$

Conversely, a subset E of R is a fundamental system of R wr to some order if it has the properties:

(i) linearly independent

(ii) every vector in R is integral linear combination of the elements of E with all coefficients of the same sign (or 0).

A suitable order is given by any v_0 in the dual space which is positive at the elements of E.
Note: Any two simple roots α_i and α_j determine a root system of rank two in the plane spanned by them. By §2.7 it is of one of the four types $A_1 \oplus A_1$,A_2,B_2,G_2. It follows easily from Proposition A (c) there that the two roots correspond to the vectors α, β of Proposition A (in some order), and that for the α_i-string of α_j one has $q = 0$ and the associated Cartan integer (written as a_{ji}) is $-p$ (Prop.A, §2.5).
It follows from (b) and (c) of Proposition A that the subgroup (N.B.: not subspace) of V generated by R (formed by the integral linear combinations of the vectors in R and called the *root lattice* $\mathcal{R}$) is a *lattice*, i.e., a free

Abelian group, discrete in V, of rank $\dim V$ and spanning V as vector space; it is generated by the basis F of V.

We interpolate an important fact.

Let $\Delta = \{\alpha, \beta, \ldots\}$ and $\Psi = \{\alpha_1, \alpha_2, \ldots, \alpha_l\}$ be the root system and fundamental system (wr to some order $>$) of our semisimple Lie algebra $\mathfrak{g}$ wr to a *CSA* $\mathfrak{h}$. To shorten the notation, we write H_i for the *fundamental coroots* H_{α_i} and X_i and X_{-i} for the root elements X_{α_i} and $X_{-\alpha_i}$ associated with the elements of Ψ. Prop.A (c), the non-vanishing of $N_{\alpha\beta}$ if $\alpha + \beta$ is a root (§2.8, Cor.B), and the relation $[X_i X_{-i}] = H_i$ imply the following:

PROPOSITION B. *The elements X_i and X_{-i} generate $\mathfrak{g}$ (as Lie algebra, i.e. under the [ ]-operation).*

We come to some new geometric concepts.

To each α in R we associate the subspace of V orthogonal to α, i.e., the set $\{\lambda \in V : \langle \alpha, \lambda \rangle = 0\}$; it is called the *singular plane* of α (of *height* 0; later we shall consider other heights) and denoted by $(\alpha, 0)$. Note $(-\alpha, 0) = (\alpha, 0)$. The Weyl reflection S_α leaves $(\alpha, 0)$ pointwise fixed and interchanges the two halfspaces of V determined by $(\alpha, 0)$. The union $\bigcup_R(\alpha, 0)$ (or $\bigcup_{R^+}(\alpha, 0)$) of all the singular planes is the *Cartan-Stiefel diagram* of R; we denote it by $D'(R)$ or just D' (more precisely this is the *infinitesimal C-S* diagram; later we will meet a global version).

The complement $V - D'$ is an open set. Its connected components are open cones in the usual sense (see Appendix), each bounded by a finite number of (parts of) singular planes $(\alpha, 0)$, its *walls*. These components are called the *Weyl chambers* of R (and their closures are the *closed Weyl chambers*). We will see below that the number of walls of any chamber is equal to the rank of R.

The $C - S$ diagram is invariant under the operation of the Weyl group $\mathcal{W}$ of R (because R is invariant and the group acts by isometries). Therefore $\mathcal{W}$ permutes the Weyl chambers. We note an important fact.

PROPOSITION C. *The Weyl group acts transitively on the set of Weyl chambers.*

Proof: Given two chambers, take a (piece-wise linear) path from the interior of one chamber to that of the other, through the interiors of the walls (i.e., avoiding the intersections of any two different singular planes); each time the path crosses a plane $(\alpha, 0)$ use the Weyl reflection S_α. (We complement this in Proposition E.) $\surd$

Let F be a fundamental system as above. The set $\{\lambda \in V : \langle \alpha_i, \lambda \rangle > 0, 1 \leq i \leq l\}$ is then a Weyl chamber C_F or C (called the *fundamental* one, for F), as follows at once from Proposition A (c): the inner product of any of its points with any element of R cannot vanish; but for each boundary point some $\langle \alpha_i, \lambda \rangle$ is 0. We also see that every Weyl chamber is linearly equivalent to the positive orthant of $\mathbb{R}^l$ (the set with all coordinates positive). More important, it follows that the Weyl group acts transitively on the set of fundamental systems, since it is transitive on the set of Weyl chambers. As a consequence, any two fundamental systems of R are congruent, and the Basic Isomorphism Theorem, Cor.C of §2.9 shows that there is an automorphism of $\mathfrak{g}$ that sends one to the other. Together with conjugacy of CSA's (Theorem F of §2.10) this yields

PROPOSITION D. *Any two fundamental systems of a complex semi-simple Lie algebra $\mathfrak{g}$ are congruent (i.e., correspond to each other under an isometry of their carrier vector spaces); in fact, there is an automorphism of $\mathfrak{g}$ sending one to the other.*

It also follows that every element α of R belongs to some fundamental system: pick a Weyl chamber that has the plane $(\alpha, 0)$ for one of its walls and lies on the positive side of the plane $\{\lambda : \langle \alpha, \lambda \rangle \geq 0\}$. The elements of R corresponding to the walls of that chamber, with suitable signs, form the desired fundamental system. To put it differently, the orbit of F under $\mathcal{W}$ is R; we have $\mathcal{W} \cdot F = R$.

Another simple consequence is the fact that $\mathcal{W}$ is generated by the Weyl reflections $S_i, 1 \geq i \geq l$, corresponding to the simple roots α_i in F: Indeed, for any two roots α and β one sees easily from geometry that the conjugate $S_\alpha \cdot S_\beta \cdot {S_\alpha}^{-1}$ is the reflection $S_{\beta'}$ with $\beta' = S_\alpha(\beta)$ (one shows that β' goes to $-\beta'$ and that any λ orthogonal to β' goes to itself, by using the analogous properties of S_α). Therefore, if the subgroup generated by the S_i contains the reflections in the walls of any given Weyl chamber, it also contains the reflections in the walls of any adjacent chamber (i.e., one obtained by reflection across a wall of the first one). Starting from the fundamental chamber we can work our way to any chamber; thus we can generate all S_α, and so all of $\mathcal{W}$.

Although we need it only much later, we prove here that the action of $\mathcal{W}$ on the set of Weyl chambers is *simply transitive.*

PROPOSITION E. *If an element of $\mathcal{W}$ leaves a Weyl chamber fixed (as a set), then it is the unit element* 1 *(or id).*

By the discussion above this is equivalent to the statement: If an element leaves a fundamental system F fixed (as a set), or leaves the positive subset R^+ fixed (as a set), then it is 1.

We first prove a lemma that expresses a basic property.

LEMMA F. *Consider α in R^+ and α_i in F, with $\alpha \neq \alpha_i$; then $S_i(\alpha)$ is also in R^+.*

(Here S_i is the Weyl reflection associated with α_i; note $S_i(\alpha_i) = -\alpha_i$.) In words: S_i sends only one positive element to a negative one, namely α_i.

Proof: By (c) of Proposition A the element α is of the form $\sum n_j\alpha_j$, with all $n_j \geq 0$ and some n_k, with $k \neq i$, different from 0. The formula $S_i(\alpha) = \alpha - a_{\alpha\alpha_i}\alpha_i$ shows that the α_k-coefficient of $S_i(\alpha)$ is still n_k and so still positive. It follows from (c) of Proposition A that $S_i(\alpha)$ is in R^+. $\surd$

For S in $\mathcal{W}$ we denote by r_S the number of positive elements of R that are sent to negative ones by S; this is called the *length* of S (wr to the given order). There is a geometric interpretation for the length: Let λ be any point in the fundamental Weyl chamber; then r_S equals the number of planes in the Cartan-Stiefel diagram that are met (and traversed) by the segment from λ to $S(\lambda)$. (N.B.: the planes $(\alpha, 0)$ and $(-\alpha, 0)$ count as the same.) Reason: We have $\langle S(\lambda), \alpha\rangle = \langle \lambda, S^{-1}(\alpha)\rangle$ (since S is an isometry); clearly we have $r_S = r_{S^{-1}}$. Since $\langle\lambda, \alpha\rangle$ is positive for all positive α, we see that $\langle S(\lambda), \alpha\rangle$ is negative for exactly r_S positive α. $\surd$

COROLLARY G. *For any S in $\mathcal{W}$ we have $(-1)^{r_S} = \det S$.*

Proof: An elementary argument shows that r_S is additive mod 2. Thus both $\det S$ and $(-1)^{r_S}$ are homomorphisms of the Weyl group into $\mathbb{Z}/2$. By Lemma F they agree on the set F of generators of $\mathcal{W}$. $\surd$

We come to the proof of Proposition E. Suppose S, with a representation $S_{i_m} \cdot S_{i_{m-1}} \cdot \quad \cdots \quad \cdot S_{i_1}$ sends F to itself. To show $S = 1$, we proceed by induction on m. For $m = 0$ we have indeed $S = 1$. With S as given we apply the reflections S_{i_1} , S_{i_2} , ... in succession to the root α_{i_1}. The first step yields $-\alpha_{i_1}$, which lies in R^-. Let S_{i_k} be the first one that brings us back to R^+ (this exists by hypothesis!). Denoting the product $S_{i_{k-1}} \cdot S_{i_{k-2}} \cdot \quad \cdots \quad \cdot S_{i_2}$ by T, we conclude from Lemma F that $T \cdot S_{i_1}(\alpha_{i_1})$ must be $-\alpha_{i_k}$, i.e., we have $T(\alpha_{i_1}) = \alpha_{i_k}$. As above, elementary geometry implies $T^{-1} \cdot S_{i_k} \cdot T = S_{i_1}$ (the left-hand side is a reflection and it sends α_{i_1} to $-\alpha_{i_1}$). We write S as $S_{i_m} \cdot \quad \cdots \quad \cdot S_{i_{k+1}} \cdot T \cdot T^{-1} \cdot S_{i_k} \cdot T \cdot S_{i_1}$, which equals then $S_{i_m} \cdot \quad \cdots \quad \cdot S_{i_{k+1}} \cdot T \cdot S_{i_1} S_{i_1} = S_{i_m} \cdot \quad \cdots \quad \cdot S_{i_{k+1}} \cdot S_{i_{k-1}} \cdot \quad \cdots \quad S_{i_2}$ (recall $S_{i_1} \cdot S_{i_1} = 1$), which is shorter by two factors; this is the induction step. $\surd$

One sees easily that Prop.E can be restated as saying: If S has a fixed point in (the interior of) a Weyl chamber, then it is the identity. We prove a consequence:

PROPOSITION H. *For any ρ in V the orbit $\mathcal{W} \cdot \rho$ under the Weyl group meets every closed Weyl chamber in exactly one point.*

(Thus the space of orbits or equivalence classes under the Weyl group can be identified with any given Weyl chamber; usually one takes the fundamental one as set of representatives for the orbits.)

We prove first that the stability group $\mathcal{W}_\rho$ of ρ, i.e., the subgroup of the Weyl group consisting of the elements that keep ρ fixed, (a) is generated by the reflections S_α for those roots α that are orthogonal to ρ, and (b) is simply transitive on the set of Weyl chambers that contain ρ in their closures.

For this purpose consider the set R' of all of those roots α for which ρ lies in the singular plane $(\alpha, 0)$, i.e., which are orthogonal to ρ. The space $((R')) = V'$ is the orthogonal complement, in V, of the intersection of the singular planes for the roots in R'; the set R' is a root system in V' and so defines Weyl chambers in V'. Their translates by ρ are the intersections of the linear variety $V' + \rho$ with those Weyl chambers of R whose closures contain ρ (let us write temporarily W_ρ for the set of these). Then the Weyl group $\mathcal{W}'$ of R' (which is a subgroup of $\mathcal{W}$ in a natural way) is transitive (in fact, simply transitive) on W_ρ. This implies $\mathcal{W}' = \mathcal{W}_\rho$ (the elements of $\mathcal{W}'$ clearly keep ρ fixed; in the other direction, $\mathcal{W}_\rho$ clearly permutes the elements of W_ρ, and using Prop.D we see that each of its elements is an element of $\mathcal{W}'$.

Prop.H follows now by counting: There are $|\mathcal{W}|/|\mathcal{W}_\rho|$ points in the orbit $\mathcal{W} \cdot \rho$, each point belongs to $|\mathcal{W}_\rho|$ closed Weyl chambers, and each closed Weyl chamber contains at least one point (by transitivity of $\mathcal{W}$). √

The number of singular planes $(\alpha, 0)$ that contain ρ is called the *degree of singularity* of ρ. Elements of V that lie on no singular plane, i.e., points in the interior of a Weyl chamber, are called *regular.*

We insert a geometric property, related to our order $>$.

PROPOSITION I. *Let λ, μ be two elements of the closed fundamental Weyl chamber $C^{\top -}$ of $\mathfrak{h}_0^\top$. Then μ lies in the convex hull of the orbit $\mathcal{W} \cdot \lambda$ of λ iff the relation $\lambda(H) \geq \mu(H)$ holds for all H in the fundamental Weyl chamber of $\mathfrak{h}_0$.*

First a lemma.

LEMMA J. *Let λ be an element of $C^{\top -}$. Then any λ' in $\mathcal{W} \cdot \lambda$ is of the form $\lambda - \sum_{\alpha > 0} c_\alpha \cdot \alpha$ with all $c_\alpha \geq 0$.*

Proof: Take a λ' in $\mathcal{W} \cdot \lambda$, different from λ. By Prop.H we know that λ' is not in $C^{\top -}$, and so there is a positive root α with $\lambda'(H_\alpha) < 0$. Thus we have $S_\alpha(\lambda'), = \lambda' - \lambda'(H_\alpha)\alpha, > \lambda'$.
After a finite number of steps we must arrive at λ itself. √

COROLLARY K. *For λ in $C^{\top -}$, S in $\mathcal{W}$ with $S\lambda \neq \lambda$, and H in C, we have $\lambda(H) > S\lambda(H) (= \lambda(S^{-1}H))$ (and $\lambda(H) \geq S\lambda(H)$ for H in C^-).*

Proof: Immediate from Lemma J, since we have $\alpha(H) > 0$, if $\alpha > 0$. $\surd$

We now prove Prop. I.

(a) Suppose $\mu = \sum_{\mathcal{W}} r_S \cdot S\lambda$ with $r_S \geq 0$ and $\sum r_S = 1$. By Cor. K we have $\mu(H) = \sum r_S S\lambda(H) \leq \sum r_S \lambda(H) = \lambda(H)$ for H in C^-.

(b) Suppose μ is not in the convex hull of $\mathcal{W} \cdot \lambda$. Then there exists H in $\mathfrak{h}_0$ with $\mu(H) > S\lambda(H)$ for all S in $\mathcal{W}$ (separation property of convex sets). By continuity of μ and of the $S\lambda$ we may take H to be regular. Then H is $T \cdot H'$ for some T in $\mathcal{W}$ and some H' in C. Now we have, using Cor. K, $\mu(H') \geq T^{-1}\mu(H') = \mu(H)$ ¿ $T\lambda(H) = \lambda(T^{-1}H) = \lambda(H')$. $\surd$

We come to the last topic of this section, the notion of *maximal* or *dominant* element of a root system (wr to the given order in V).

First an important definition: An element α of R^+ is called *extreme* or *highest* , if $\alpha + \beta$ is not a root for any positive root β.

(Actually this is equivalent to requiring that $\alpha + \alpha_i$ is not a root for any fundamental root α_i. Writing β as sum of a positive and a fundamental root if it is not fundamental itself (Prop.A (d)), one reduces this to the following: If α,β,γ,$\alpha+\beta$, and $\alpha+\beta+\gamma$ are in R, then at least one of $\alpha+\gamma$ and $\beta+\gamma$ is also a root. This in turn follows easily from the Jacobi identity for X_α,X_β,X_γ and the fact that $N_{\alpha\beta}$ is different from 0 iff $\alpha+\beta$ is a root.)

PROPOSITION L. *Let R be a simple root system (with order and fundamental system as above). Then there exists a unique extreme root μ, the maximal or dominant element of R; μ is the unique maximal (wr to $>$) root and lies in the fundamental Weyl chamber. Moreover, with μ expressed as $\sum m_j\alpha_j$ and an arbitrary root β as $\sum b_j\alpha_j$ the inequalities $m_i \geq b_i$ hold for $1 \leq i \leq l$; in particular, the m_i are all positive.*

Proof: Let $\alpha = \sum a_i\alpha_i$ be an extreme root. We have $\langle\alpha, \alpha_i\rangle \geq 0$ for all i, by extremeness (otherwise $\alpha + \alpha_i$ would be in R by Prop.B of §2.7); thus α is in the fundamental Weyl chamber.

Next we show that all a_i are positive. They are non-negative to begin with (α is in R^+). If some a_k is 0, then we have $\langle\alpha, \alpha_k\rangle \leq 0$,since $\langle\alpha_i, \alpha_k\rangle \leq 0$ for $i \neq k$. Together with the previous inequality this gives $\langle\alpha, \alpha_k\rangle = 0$; and this in turn implies $\langle\alpha_i, \alpha_k\rangle = 0$ for all the i with $a_i \neq 0$ and all the k with $a_k = 0$. Thus F would split into two non-empty, mutually orthogonal sub systems F' and F''. But then R would split in a similar way, contradicting its simplicity: As noted after Prop. D, R is the orbit of F under the Weyl group of F, and this Weyl group is of course the direct product of the Weyl groups of F' and F'', acting in the obvious way.

Let now α and β be two extreme elements. First we have $\langle\alpha, \beta\rangle \geq 0$; otherwise $\alpha + \beta$ is in R. Since $\langle\alpha_i, \beta\rangle \geq 0$ and $a_i > 0$ for all i, the relation $\langle\alpha, \beta\rangle = 0$ would imply that all $\langle\alpha_i, \beta\rangle$ vanish; but that would mean $\beta = 0$. Thus $\langle\alpha, \beta\rangle > 0$, and so $\alpha - \beta$ is in R (or is 0). Say it is in R^+; then we get the impossible relation $\alpha = \beta + (\alpha - \beta)$. This means $\alpha = \beta$, and uniqueness of the extreme μ is established. Maximality (and uniqueness of maximal elements) follows from the obvious fact that maximal elements (which exist by finiteness) are extreme.

2.12 Fundamental systems

Fundamental systems of root systems are important enough to warrant a definition:

DEFINITION A. *An (abstract) fundamental system is a non-empty, finite, linearly independent subset $F = \{\alpha_1, \alpha_2, ..., \alpha_l\}$ of a Euclidean space (=real vector space with positive definite inner product $\langle \cdot, \cdot \rangle$) such that for any α_i and α_j in F the value $2\langle \alpha_i, \alpha_j \rangle / \langle \alpha_j, \alpha_j \rangle = a_{ij}$ is a non-positive integer.*

The a_{ij} are the *Cartan integers* of F; they form the *Cartan matrix* $A = [a_{ij}]$. One sees as in §2.7 that only the values $0, -1, -2, -3$ can occur for $i \neq j$ and that the table of §2.7 applies to any two vectors in F. (In the literature one also finds a_{ji} for our a_{ij}, i.e., the indices are reversed.)

Usually one assumes $((F)) = V$.

Equivalence of fundamental systems is defined, as for root systems, as a bijection induced by a similarity of the ambient Euclidean spaces. There is again a Weyl group $\mathcal{W}$, generated by the reflections S of V in the hyperplanes orthogonal to the α_i. $\mathcal{W}$ is again finite: The formula $S_i(\alpha_j) = \alpha_j - a_{ji}\alpha_i$ shows that each S_i leaves the lattice $\mathcal{R}$ generated by F invariant; since the elements of $\mathcal{W}$ are isometries of V, there are only finitely many possibilities for what they can do to the vectors in F.

There is the notion of *decomposable* fundamental system: union of two non-empty mutually orthogonal subsets. Every fundamental system splits uniquely into mutually orthogonal *simple* (= not decomposable) ones.

In §2.11 we associated with every root system R a fundamental system F contained in it, unique up to an operation of the Weyl group of R. F in turn determines R: First, since the reflections S_i attached to the elements of F generate the Weyl group of R (as we saw), the Weyl groups of R and F are identical. Second, we showed (in effect) that the orbit $\mathcal{W} \cdot F$, the set of the $S(\alpha_i)$ with S in $\mathcal{W}$ and α_i in F, is R.

The main conclusion from all this for us is that in order to construct all root systems it is enough to construct all fundamental systems. This turns out to be quite easy; we do it in the next section.

To complete the picture we should also show that every (abstract) fundamental system comes from a root system. One way to do this is to construct all possible (abstract) fundamental systems (we do this in the next section), and to verify the property for each case (we will write down the root systems explicitly).

There is also a general way of proceeding: The root system would have to be, of course, the orbit $\mathcal{W} \cdot F$ of F under its Weyl group. We first have to show that this set R is indeed a root system. We prove properties $(i)'$, $(ii)'$ and (iii) of §2.6. First, the Weyl group of R is again identical with that of F, since for any $\alpha = S(\alpha_i)$ we have $S_\alpha = S \cdot S_i \cdot S^{-1}$, and so S_α is in the Weyl group of F. It follows that R is invariant under its Weyl group, i.e., $(ii)'$ holds. Next, for any β in R we have $S_i(\beta) - \beta = n \cdot \alpha_i$ with integral n (the left-hand side is in the lattice $\mathcal{R}$ and is a real multiple of α_i, and α_i is a primitive element of the lattice). Applying S and recalling the relation $S_\alpha = S \cdot S_i \cdot S^{-1}$, we get $S_\alpha \cdot S(\beta) - S(\beta) = nS(\alpha_i) = n_\alpha$. This proves $(i)'$,since $S(\beta)$ runs over all of R as β does. Finally, for (iii) we note that α is also a primitive element of the lattice, since S is invertible.

We still have to prove that the given F is a fundamental system of the root system $R = \mathcal{W} \cdot F$ defined by it. That is not quite so obvious. It amounts to showing that the fundamental Weyl chamber C_F of F, i.e. the set $\{\lambda : \langle \lambda, \alpha_i \rangle > 0, 1 \leq i \leq l\}$ is identical with the corresponding chamber C_R of R (clearly we have $C_R \subset C_F$ anyway), or that the $\mathcal{W}$-transforms of C_F are pairwise disjoint. We proceed by induction on $\dim V$. The situation is trivial for $\dim = 0$, and also for $\dim = 1$; in the latter case F consists of one vector α, with $R = \{\alpha, -\alpha\}$, $\mathcal{W} = \{id, -id\}$, $C_F = C_R = \{t\alpha : t > 0\}$. The case $\dim = 2$ is a bit exceptional; we have in effect considered it in §2.7, when we constructed all root systems of rank 2. According to the table there, there are four possibilities for F, and one easily verifies our claim for each case.

Now the induction step, assuming $l = \dim V > 2$. Let Σ be the unit sphere in V. Choose r with $1 \leq r \leq l$, and let v be a point of Σ in the closure of C_F that lies on exactly r singular planes $(\alpha_i, 0)$, i.e. that is orthogonal to r of the elements of F. These r elements form a fundamental system F_v, whose Weyl group $\mathcal{W}_v$ is a subgroup of $\mathcal{W}$. Our induction assumption holds for this system. This means that the $\mathcal{W}_v$-transforms of the fundamental chamber C_F fit together around v without overlap. We interpret this on Σ: Let D denote the intersection of Σ with the closure of C_F; this is a (convex) spherical cell. Then the $\mathcal{W}_v$-transforms of D will fit together around v, meeting only in boundary points and filling out a neighborhood of v on Σ. We form a cell complex by taking all the transforms of D by the elements of $\mathcal{W}$ and attaching them to each other as indicated by the groups $\mathcal{W}_v$ above, at their faces of codimension $r, 1 \leq r \leq l-1$. The fact just noted about the $\mathcal{W}_v$-transforms filling out a neighborhood means that the obvious map of our cell complex onto Σ is a covering in the usual topological sense (each point in Σ has an "evenly covered" neighborhood). It is well known that the sphere Σ has only trivial coverings for $l - 1 > 1$. This means that our map is bijective, i.e. that the transforms $S \cdot D$, with S running over $\mathcal{W}$, have no interior points in common and simply cover Σ. Clearly this proves our claim, that the fundamental chamber of F is also a chamber of the root system $R = \mathcal{W} \cdot F$, and that F is a fundamental system for R. $\surd$

2.13 Classification of fundamental systems

Let $F = \{\alpha_1, \alpha_2, \ldots, \alpha_l\}$ be a fundamental system (in a Euclidean space V). To F one associates a "diagram" , a weighted graph, the *Dynkin diagram*, as follows: To each vector α_i is associated a vertex or 0-cell, provided with the *weight* $\langle \alpha_i, \alpha_i \rangle$ or $|\alpha_i|^2$ (usually written above the vertex); for any two different vertices α_i and α_j the corresponding vertices are connected by $a_{ij} \cdot a_{ji} = |a_{ij}|$ $(= 0, 1, 2, 3)$ edges or 1-cells. In particular, if $\langle \alpha_i, \alpha_j \rangle = 0$, then there is no edge. In the case of two or three edges, one often adds an arrow, pointing from the higher to the lower weight (from the longer to the shorter vector).

(Similar diagrams had been introduced by Coxeter earlier.)

For a *connected* (in the obvious sense) Dynkin diagram the weights are clearly determined (up to a common factor) by the graph (with its arrows), since the number of edges plus direction of the arrow determines the ratio of the weights. The Dynkin diagram (with weights up to a common factor) and the Cartan matrix $A = [a_{ij}]$ determine each other; the arrows are given by the fact that $|a_{ij}|$ (assumed not 0) is greater than 1 iff $|\alpha_i|$ is greater than $|\alpha_j|$.

The diagram (with the weights) determines F up to congruence : First one can find the a_{ij}, since of a_{ij} and a_{ji} one is equal to -1, and the arrow determines which one; then from the a_{ij} and the $\langle \alpha_i, \alpha_i \rangle$ one can find all $\langle \alpha_i, \alpha_j \rangle$.

There is of course the notion of abstract Dynkin diagram, i.e., a weighted diagram of this kind, but without a fundamental system in the background. Given such a diagram, one can try to construct a fundamental system from which it is derived by the obvious device of introducing the vector space V with the vertices α_i of the diagram as basis and the "inner product" $\langle \cdot, \cdot \rangle$ determined by the $\langle \alpha_i, \alpha_j \rangle$ as read off from the diagram; this will succeed precisely if the form $\langle \cdot, \cdot \rangle$ turns out positive definite.

The Dynkin diagram of a fundamental system F is connected iff F is simple; in general the connected components of a diagram correspond to the simple constituents of F. A connected diagram with its arrows, but without its weights, determines the fundamental system up to equivalence (= similarity), since it determines the norms of the vectors (or the weights) up to a common factor. One often normalizes the systems by assuming the smallest weight to be 1. It turns out to be quite simple to construct all possible fundamental systems in terms of their Dynkin diagrams.

THEOREM A. *There exist (up to equivalence) exactly the following simple fundamental systems (described by their Dynkin diagrams) :*

Name	Diagram	Rank
A_l		$l = 1, 2, 3, \dots$
B_l		$l = 2, 3, 4, \dots$
C_l		$l = 3, 4, 5, \dots$
D_l		$l = 4, 5, 6, \dots$
G_2		$l = 2$
F_4		$l = 4$
E_6	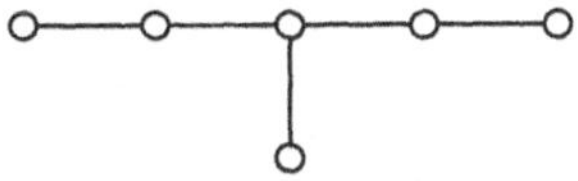	$l = 6$
E_7	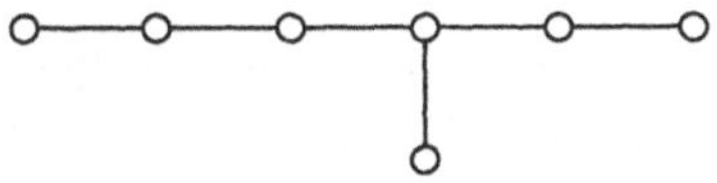	$l = 7$
E_8	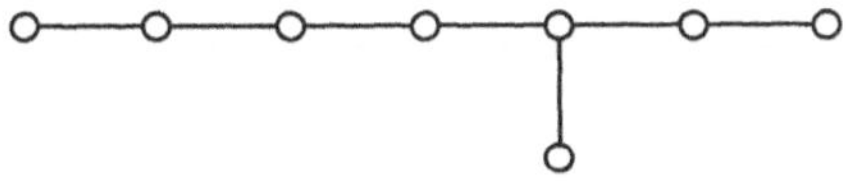	$l = 8$

The diagrams of the classes A_l, B_l, C_l, D_l (which depend on an integral parameter) are called the *four big classes* or the *classical diagrams*; the diagrams G_2, F_4, E_6, E_7, E_8 are the *five exceptional diagrams.* Same nomenclature for the corresponding fundamental systems.

We comment on the restrictions on l for the classical types; they are meant to avoid "double exposure": B_l is supposed to "end" with B_2

on its right; this requires $l \geq 2$. Put differently, proceeding formally with $l = 1$ would give B_1 as a single vertex – which would be identical with A_1.

Next, C_2 is the same diagram as B_2 (only differently situated); thus one requires $l \geq 3$ for the class C_l.

Finally D_l: Here D_3 is identical with A_3. We can interpret D_2 as the "right end" of the general D_l-diagram, consisting of two vertices and no edge; it is thus decomposable, and represents in fact the system $A_1 \oplus A_1$ (or $B_1 \oplus B_1$). D_1 could be interpreted as the empty diagram (which we didn't allow earlier); this "is" the Dynkin diagram for a one-dimensional (Abelian) Lie algebra (there are no roots). All this makes good sense in terms of the so-called *accidental isomorphisms* between certain low-dimensional classical Lie algebras and groups (see [3], also §1.1).

We note that the diagrams A_l (for $l > 1$), D_l, and E_6 have obvious *self-equivalences* (*automorphisms*) : For A_l and E_6 reversal of the horizontal arrangement, for D_l switching of the two vertices on the right. The diagram D_4 shows an exceptional behavior: it permits the full symmetric group on three objects (the endpoints) as group of automorphisms. This will be reflected in automorphisms of the corresponding Lie algebras.

For the proof of Theorem A we will construct all possible (connected) diagrams with positive form $\langle \cdot, \cdot \rangle$ by simple geometric arguments. The proof will be broken into a number of small steps. We will be using slightly undefined notions such as *subdiagram* (some of the vertices of and some of the edges connecting them in a larger diagram). For any α_i in F we write v_i for the normalized vectors $\alpha_i/|\alpha_i|$. Thus corresponding to the "basic links"

we have respectively $\langle v_i, v_j \rangle = -1/2, -1/\sqrt{2}, -\sqrt{3}/2$.

1) The diagram G_2 is not subdiagram of any larger diagram (with positive form $\langle \cdot, \cdot \rangle$): Otherwise we find a subdiagram

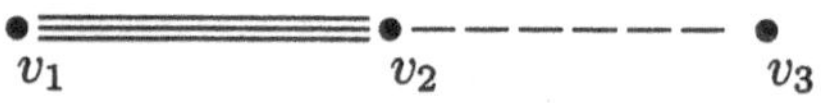

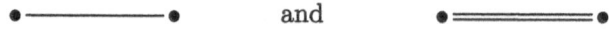

with the arrow in the G_2-part going either way and the other part one of the three basic links. This gives three vectors v_1, v_2, v_3 with $\langle v_1, v_2 \rangle = -\sqrt{3}/2$, $\langle v_1, v_3 \rangle \leq 0$, $\langle v_2, v_3 \rangle \leq -1/2$. (For the second inequality note that in the larger diagram there could be $0, 1, 2$, or 3 edges from v_1 to v_3.) For $\alpha = \sqrt{3}v_1 + 2v_2 + v_3$ we compute $\langle \alpha, \alpha \rangle \leq 0$ (we use here and below, without further comment, the fact that all $\langle v_i, v_j \rangle$ for $i \neq j$ are ≤ 0). But this contradicts positive definiteness of $\langle \cdot, \cdot \rangle$.

From now on we consider only diagrams without G_2 as subdiagram, i.e., only diagrams made up of the basic links

 and 

2) A diagram can contain B_2 only once as subdiagram: Otherwise there is a subdiagram of the type

Let $v_1, v_2, \ldots$ be the corresponding vectors (from left to right) and put $\alpha = 1/\sqrt{2}v_1 + v_2 + \ldots + v_{t-1} + 1/\sqrt{2}v_t$. One computes $\langle \alpha, \alpha \rangle \leq 0$ (note again that there might be additional edges between some of the vertices in

the big diagram). This again contradicts positive definiteness of the inner product.

3) There is no closed polygon containing B_2 in a diagram: Otherwise on going around the polygon the weight would change exactly once, by a factor 2, manifestly impossible.

4) If there is a B_2, then there is no branchpoint: Otherwise there would be a subdiagram

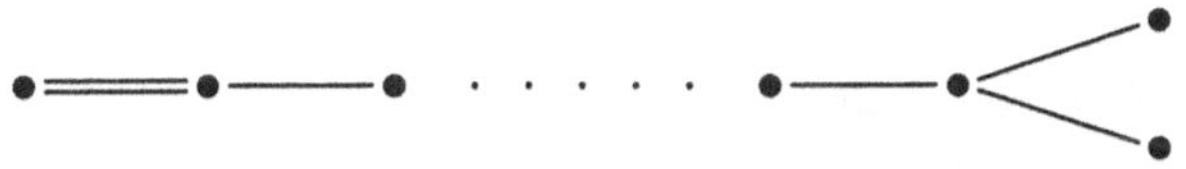

Let $v_1, v_2, \ldots, v_t$ be the vectors, in order from the left, with v_{t-1} and v_t the two ends at the right. Put $\alpha = 1/\sqrt{2}v_1 + v_2 + \ldots + v_{t-2} + 1/2(v_{t-1} + v_t)$, and verify $\langle \alpha, \alpha \rangle \leq 0$; contradiction.

5) The diagram

does not occur as subdiagram.

Reason: Put $\alpha = \sqrt{2}v_1 + 2\sqrt{2}v_2 + 3v_3 + 2v_4 + v_5$, and verify $\langle \alpha, \alpha \rangle \leq 0$.

From 2) to 5) we conclude that diagrams containing B_2 must be of the types B_l, C_l, F_4 listed in Theorem A. Therefore from now on we consider only diagrams containing neither G_2 nor B_2, i.e., made up of A_2 only.

6) There are no closed polygons in the diagram. (The diagram is a tree.)

Otherwise, with $v_1, v_2, \ldots, v_t$ the vectors around the circuit, one computes that $\alpha = \sum v_i$ has $\langle \alpha, \alpha \rangle \leq 0$.

7) There are at most three endpoints (and therefore at most one branchpoint).

Otherwise there is a subdiagram

(The horizontal part might be "empty".) Let $v_1, \ldots, v_t$ be the vectors, with v_1 and v_2 at the left ends and v_{t-1} and v_t at the right ends. Then $\alpha = 1/2(v_1 + v_2) + v_3 + \ldots + v_{t-2} + 1/2(v_{t-1} + v_t)$ has $\langle \alpha, \alpha \rangle \leq 0$.

8) If there is a branchpoint, then one of the branches has length one.

Otherwise there is a subdiagram

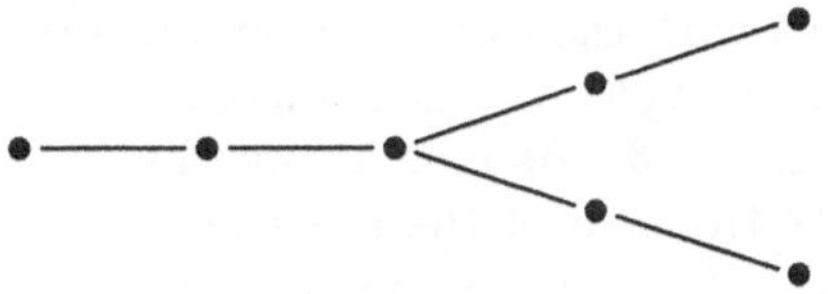

Let v_1 be the center, v_2, v_3, v_4 adjacent to it, and v_5, v_6, v_7 the endpoints. Then $\alpha = 3v_1 + 2(v_2 + v_3 + v_4) + v_5 + v_6 + v_7$ has $\langle \alpha, \alpha \rangle \leq 0$.

9) The diagram

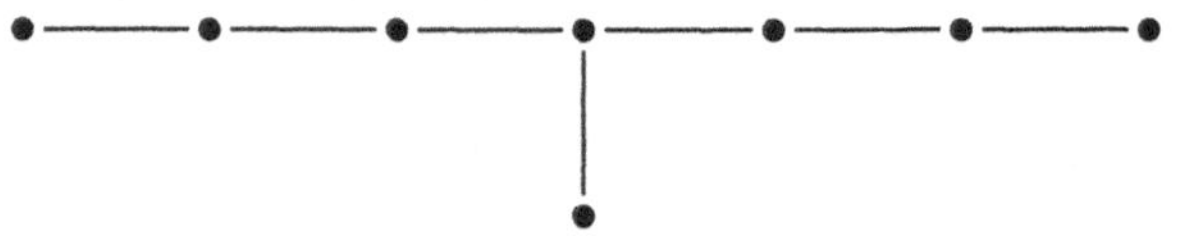

is impossible as subdiagram.

Let $v_1, \ldots, v_7$ be the vectors on the horizontal, v_8 the one below. Then $\alpha = v_1 + 2v_2 + 3v_3 + 4v_4 + 3v_5 + 2v_6 + v_7 + 2v_8$ has $\langle \alpha, \alpha \rangle \leq 0$.

10) The diagram

is impossible as subdiagram.

With the analogous numbering put $\alpha = v_1 + 2v_2 + 3v_3 + 4v_4 + 5v_5 + 6v_6 + 4v_7 + 2v_8 + 3v_9$ and verify $\langle \alpha, \alpha \rangle \leq 0$.

From 6) to 10) it follows easily that diagrams with all links of type A_2 must be A_l, D_l, E_6, E_7, or E_8 of Theorem A. √

As noted before, we still have to show that the diagrams listed in Theorem A are Dynkin diagrams of fundamental systems, i.e., that the corresponding quadratic form is positive definite. (We verify that this is so in the next section, where we will write down the fundamental systems and root systems for each case.) As an example we look at F_4. The quadratic form works out to ${x_1}^2 + {x_2}^2 + 2{x_3}^2 + 2{x_4}^2 - x_1x_2 - 2x_2x_3 - 2x_3x_4$. By completing squares this can be written as $(x_1 - 1/2x_2)^2 + 1/4(x_2 - 4/3x_3)^2 + 2/3(x_3 - 3/2x_4)^2 + 1/2{x_4}^2$. √

We comment on how the vectors α with $\langle \alpha, \alpha \rangle \leq 0$ were constructed above: Recursively the coefficients of the v_i are so chosen that the norm square of each vector cancels the sum of the inner products with the adjacent (in the subdiagram) vectors. (Any additional links in the original diagram contribute non-positive amounts.) Take 5) as an example: We start with v_5. The factor r of v_4 is determined from the relation $\langle v_5, v_5 \rangle + \langle rv_4, v_4 \rangle = 0$; with the rule noted just before 1) this gives $r = 2$. The next equation, involving the coefficient s of v_3, is $\langle 2v_4, 2v_4 \rangle + \langle 2v_4, v_5 \rangle + \langle 2v_4, sv_3 \rangle = 0$, yielding $s = 3$. (As long as only links A_2 occur, the rule is: each coefficient is 1/2 the sum of the adjacent ones.) The factor t of v_2 comes from $\langle 3v_3, 3v_3 \rangle + \langle 3v_3, 2v_4 \rangle + \langle 3v_3, tv_2 \rangle = 0$ as $t = 2\sqrt{2}$. The next step, $\langle 2\sqrt{2}v_2, 2\sqrt{2}v_2\rangle + \langle 2\sqrt{2}v_2, 3v_3\rangle + \langle 2\sqrt{2}v_2, uv_1\rangle = 0$ gives the factor u of v_1 as $\sqrt{2}$. This happens to be one-half the factor of v_2; this "accident" is responsible for $\langle\alpha,\alpha\rangle \le 0$ (the sum of the squares cancels twice the sum of the relevant inner products).

In case 9) we start with v_1 and work our way to v_4; the factor of v_8 is one-half that of v_4; then we find v_5 etc.

2.14 The simple Lie algebras

The next step in our program is to show that each of abstract Dynkin diagrams found in §2.13 comes from the fundamental system for the root system of some semisimple Lie algebra. There are several approaches to this problem.

The most direct approach (Serre) uses the entries a_{ij} of the Cartan matrix A (with $1 \le i,j \le l$), and defines the Lie algebra by generators and relations: There are $3l$ generators e_i, f_i, h_i with $1 \le i \le l$ (corresponding to the elements X_i, X_{-i}, H_i of $\mathfrak{g}$ introduced in §2.11); the relations are

$$(1) \qquad [h_i h_j] = 0,$$

$$(2) \qquad [h_i e_j] = a_{ji}e_j \text{and} [h_i f_j] = -a_{ji}f_j,$$

$$(3) \qquad [e_i f_j] = 0,$$

$$(4) \qquad [e_i[e_i[...[e_ie_j]...] = 0 \text{ for } -a_{ji}+1 \text{ factors } e_i$$

$$(5) \qquad [f_i[f_i...[f_if_j]...] = 0 \text{ for } -a_{ji}+1 \text{ factors } f_i$$

One proves that this is a (finite dimensional!) semisimple Lie algebra with the correct root and fundamental system. The h_i form a Cartan sub Lie algebra. (See [24].)

Another approach (Tits, [23]) uses the relations between the a_{ij} (or equivalently the strings) and the $N_{\alpha\beta}$ of §2.8, 2.9 to show that the $N_{\alpha\beta}$ can be so chosen (recall they are determined up to some signs) that the result is in fact a Lie algebra, with the correct root system.

We shall not reproduce these arguments here, but shall follow the traditional path of Killing and Cartan of simply writing down the necessary Lie algebras. That turns out to be easy for the four classical classes. For the five exceptional we write down the root system, but do not enter into the rather long verification of the fact that there is a Lie algebra behind the root system.

We state the main result.

THEOREM A. *Assigning to each complex semisimple Lie algebra the Dynkin diagram of the root system of a Cartan sub Lie algebra sets up a bijection between the set of (isomorphism classes of) such Lie algebras and the set of (equivalence classes of) abstract fundamental systems. In particular, the simple Lie algebras correspond to the simple diagrams, listed in Theorem A of §2.13, and are given by the following table :*

Name	Description	Rank	Dimension
A_l	$\mathfrak{sl}(l+1, \mathbb{C})$	$l = 1, 2, \ldots$	$l(l+2)$
B_l	$\mathfrak{o}(2l+1, \mathbb{C})$	$l = 2, 3, \ldots$	$l(2l+1)$
C_l	$\mathfrak{sp}(l, \mathbb{C})$	$l = 3, 4, \ldots$	$l(2l+1)$
D_l	$\mathfrak{o}(2l, \mathbb{C})$	$l = 4, 5, \ldots$	$l(2l-1)$
G_2	–	2	14
F_4	–	4	52
E_6	–	6	78
E_7	–	7	133
E_8	–	8	248

Corresponding Lie algebras and Dynkin diagrams are denoted by the same symbol. A_l, B_l, C_l, D_l are the *classical Lie algebras*; G_2, F_4, E_6, E_7, E_8 are the *five exceptional* ones (just as for the diagrams). (We note that in using these classical names we are deviating from our convention on notation, §1.1.) It is clear from the earlier discussion and the comments above on the exceptional cases that all that remains to be done here is to verify that the classical Lie algebras have the correct fundamental systems or Dynkin diagrams. We proceed to do this. All these Lie algebras are sub Lie algebras of $\mathfrak{gl}(n, \mathbb{C})$ for appropriate n, i.e., their elements are matrices of the appropriate size. We write E_{ij} for the usual matrix "unit" with 1 as ij-entry and 0 everywhere else. We use the standard basis vectors e_i of $\mathbb{R}$ and $\mathbb{C}$ and the standard linear functionals ω_i (see Appendix). In each case we shall display an Abelian sub Lie algebra $\mathfrak{h}$, which is in fact a CSA, and the corresponding roots, fundamental system and (for the classical cases) root elements, and also the fundamental coroots and the Cartan matrix; the proof that the displayed objects are what they are claimed to be, and that the Lie algebra itself is semisimple, will mostly be omitted.

As for the dimensions in the table above: It is clear from the general structure that the dimension of a semisimple Lie algebra is equal to the sum of rank and number of roots.

1) A_l.

For $\mathfrak{sl}(l+1, \mathbb{C})$ one can take as CSA $\mathfrak{h}$ the space of all diagonal matrices $H = \mathrm{diag}(a_1, a_2, \ldots, a_{l+1})$ (with $\sum a_i = 0$). We treat $\mathfrak{h}$ in the obvious way as the subspace of $\mathbb{C}^{l+1}$ on which $\sum \omega_i$ vanishes. One computes $[HE_{ij}] = (a_i - a_j)E_{ij}$; thus the linear functions $\alpha_{ij} = \omega_i - \omega_j$, for $i \neq j$, are the roots and the E_{ij}, for $i \neq j$, are the root elements. $\mathfrak{h}_0$ is obtained by taking all a_i real; it is thus $\mathfrak{h} \cap \mathbb{R}^{l+1}$. We define the order in $\mathfrak{h}_0^\top$ through some (arbitrarily chosen) H_0 in $\mathfrak{h}_0$ with $a_1 > a_2 > \ldots > a_{l+1}$. The positive roots are then the α_{ij} with $i < j$. The fundamental system consists of $\alpha_{12}, \alpha_{23}, \alpha_{34}, \ldots, \alpha_{l,l+1}$; for $i < j$ we have $\alpha_{ij} = \alpha_{i,i+1} + \alpha_{i+1,i+2} + \cdots + \alpha_{j-1,j}$. The fundamental Weyl chamber consists of the H with $a_1 > a_2 > \cdots > a_{l+1}$. The maximal root is $\alpha_{12} + \alpha_{23} + \cdots + \alpha_{l,l+1}, = \omega_1 - \omega_{l+1}$.

The only way to form non-trivial strings of roots is to add two "adjacent" roots: α_{ij} and α_{kl} with either $j = k$ or $i = l$. This means that for two adjacent fundamental roots we have $q = 0$ and $p = 1$ (in the notation of §2.5), so that then the Cartan integer is -1, and that non-adjacent fundamental roots are orthogonal to each other. Thus the Dynkin diagram is

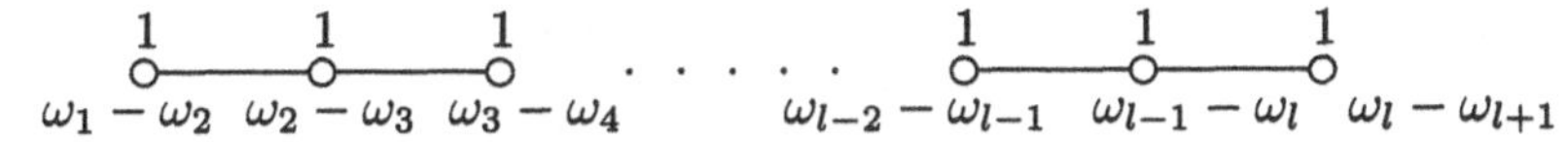

The fundamental coroots are $H_1 = e_1 - e_2, H_2 = e_2 - e_3, \ldots, H_l = e_l - e_{l+1}$.

One verifies that the bracket of two root elements E_{ij} and E_{jk} is non-zero exactly if the sum of the two roots α_{ij} and α_{kl} is again a root (meaning $j = k$ or $i = l$), in accordance with our general theory. In fact, the opposite view of structure theory is possibly sounder: the general semisimple Lie algebra has a structure similar to that of $\mathfrak{sl}(n, \mathbb{C})$, as exhibited above.

As for simplicity of A_l, it is elementary that there are no ideals: starting from any non-zero element it is easy, by taking appropriate brackets, always going up in the order, to produce the element $E_{1,l+1}$, and then, by taking further brackets, all E_{ij} and all H (note $[E_{ij}E_{ji}] = E_{ii} - E_{jj} = e_i - e_j$; these elements span $\mathfrak{h}$).

$\mathfrak{h}$ is a Cartan sub Lie algebra since it is nilpotent (even Abelian) and clearly equals its own normalizer. The Killing form on $\mathfrak{h}$ (sum of the squares of all roots) is, up to a factor, the Pythagorean expression $\sum_1^{l+1} {\omega_i}^2$. (Note that because of $\sum \omega_i = 0$ we have $\sum_{i \neq j} \omega_i \omega_j = -\sum {\omega_i}^2$.)

As for the Weyl group $\mathcal{W}$, the reflection S_{12}, corresponding to the root α_{12}, clearly consists in the interchange of the coordinates a_1 and a_2 of any H. One concludes that $\mathcal{W}$ consists of all permutations of the coordinate axes, and is thus the full symmetric group on $l + 1$ elements.

The Cartan matrix has 2's on the main diagonal, and -1's on the two diagonals on either side of the main one.

In the remaining cases we shall give less detail.

2) B_l.

For $\mathfrak{o}(2l+1, \mathbb{C})$, the orthogonal Lie algebra in an odd number of variables, we take instead of the usual quadratic form $\sum x_i{}^2$ the variant $x_0{}^2+2(x_1x_2+x_3x_4+\ldots+x_{2l-1}x_{2l})$, which leads to somewhat simpler formulae; i.e., with $P = \begin{bmatrix} 0 & 1 \\ 1 & 0 \end{bmatrix}$ and $K = \operatorname{diag}(1, P, \ldots, P) = E_{00} + E_{12} + E_{21} + E_{34} + E_{43} + \ldots + E_{2l-1,2l} + E_{2l,2l-1}$, we take our Lie algebra to be the set of matrices A that satisfy $A^\top K + KA = 0$. For $\mathfrak{h}$ we take the sub Lie algebra of diagonal matrices; they are of the form $H = \operatorname{diag}(0, a_1, -a_1, a_2, -a_2, \ldots, a_l, -a_l)$. We treat $\mathfrak{h}$ as $\mathbb{C}^l$, with H corresponding to $(a_1, a_2, \ldots, a_l)$; the real subspace $\mathbb{R}^l$ is $\mathfrak{h}_0$.

The roots and root elements are then described in the following table.

Roots		Root elements
ω_i	*for* $1 \le i \le l$	$\sqrt{2}(E_{2i-1,0} - E_{0,2i})$
$-\omega_i$	"	$\sqrt{2}(E_{0,2i-1} - E_{2i,0})$
$\omega_i - \omega_j$	$i \neq j$	$E_{2i-1,2j-1} - E_{2j,2i}$
$\omega_i + \omega_j$	$i < j$	$E_{2j-1,2i} - E_{2i-1,2j}$
$-\omega_i - \omega_j$	"	$E_{2i,2j-1} - E_{2j,2i-1}$

The order in $\mathfrak{h}_0^\top$ is defined by some H_0 with $a_1 > a_2 > \ldots > a_l > 0$; the positive roots are the ω_i and the $\omega_i \pm \omega_j$ with $i < j$.
The fundamental system is $\{\omega_1 - \omega_2, \omega_2 - \omega_3, \ldots, \omega_{l-1} - \omega_l, \omega_l\}$; one verifies that every positive root is sum of some of these.
The fundamental Weyl chamber is given by $a_1 > a_2 > \ldots > a_l > 0$.
The maximal root is $\omega_1 + \omega_2 = \omega_1 - \omega_2 + 2(\omega_2 - \omega_3 + \ldots + \omega_{l-1} - \omega_l + \omega_l)$.

Now the diagram: for the first $l-1$ fundamental roots we can form strings only by adding adjacent roots; this means that we have links of type A_2 between adjacent roots. For the last pair, $\omega_{l-1} - \omega_l$ and ω_l, one cannot subtract ω_l, but one can add it twice to $\omega_{l-1} - \omega_l$; thus the Cartan integer is -2 and there is a link of type A_2 with the arrow going from $\omega_{l-1} - \omega_l$ to ω_l. The Dynkin diagram is then

The Killing form is again $\sum \omega_i{}^2$, up to a factor, as easily verified.

The Weyl group contains the interchange of any two axes (Weyl reflection corresponding to $\omega_i - \omega_j$) and the change of any one coordinate into the negative (corresponding to the root ω_i). Thus it can be considered as the group of all permutations and sign changes on l variables; the order is $2l \cdot l!$.

Fundamental coroots: $H_1 = e_1 - e_2, \ldots, H_{l-1} = e_{l-1} - e_l, H_l = 2e_l$.

The Cartan matrix differs from that of A_l only by having -2 as $(l-1,l)$-entry.

(If we use $\sum_0^{2l} {x_i}^2$ as the basic quadratic form, then the relevant Cartan sub Lie algebra consists of the matrices of the form

$$\text{diag}(0, a_1 J_1, a_2 J_1, \ldots, a_l J_1),$$

with the usual matrix J_1, and the a_i in $\mathbb{C}$, purely imaginary for $\mathfrak{h}_0$.)

3) C_l.

$\mathfrak{sp}(l,\mathbb{C})$ consists of the $2l \times 2l$ matrices M satisfying $M^\top J + JM = 0$ (see §1.1 for J). We let $\mathfrak{h}$ be the set of matrices

$$H = \text{diag}(a_1, -a_1, a_2, -a_2, \ldots, a_l, -a_l),$$

setting up the obvious isomorphism with $\mathbb{C}^l$. As before we have $\mathfrak{h}_0 = \mathbb{R}^l$.

Roots		Root elements
$\omega_i - \omega_j$	$i \neq j$	$E_{2i-1,2j-1} - E_{2j,2i}$
$\omega_i + \omega_j$	$i < j$	$E_{2i-1,2j} + E_{2j-1,2i}$
$-\omega_i - \omega_j$	$i < j$	$E_{2i,2j-1} + E_{2j,2i-1}$
$2\omega_i$		$E_{2i-1,2i}$
$-2\omega_i$		$E_{2i,2i-1}$

Order in $\mathfrak{h}_0^\top$ defined by $H_0 = (l, l-1, \ldots, 1)$.
Positive roots: $\omega_i - \omega_j$ and $\omega_i + \omega_j$ with $i < j, 2\omega_i$.
Fundamental system: $\omega_1 - \omega_2, \omega_2 - \omega_3, \ldots, \omega_{l-1} - \omega_l, 2\omega_l$.
Fundamental Weyl chamber: $a_1 > a_2 > \ldots > a_l$.
Maximal root: $2\omega_1, = 2(\omega_1 - \omega_2 + \omega_2 - \omega_3 + \ldots + \omega_{l-1} - \omega_l) + 2\omega_l$.

For the first $l-1$ fundamental roots there is an A_2-link from each to the next. For the last pair, the $(\omega_{l-1} - \omega_l)$-string of $2\omega_l$ has $q = 0$ and $p = 2$. Thus the Dynkin diagram is

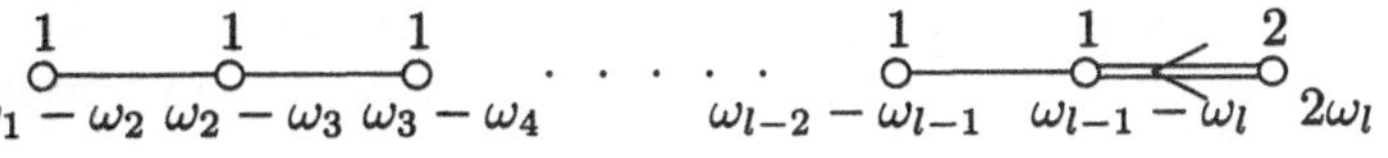

The Killing form is again $k \cdot \sum {\omega_i}^2$. We note that B_l and C_l have the same infinitesimal diagram and the same Weyl group (but the roots are not the same: C_l has $\pm 2\omega_i$ where B_l has $\pm\omega_i$).

Fundamental coroots: $H_1 = e_1 - e_2, \ldots, H_{l-1} = e_{l-1} - e_l, H_l = e_l$.

The Cartan matrix is the transpose of that for B_l.

4) D_l.

For $\mathfrak{o}(2l, \mathbb{C})$, the orthogonal Lie algebra in an even number of variables, we take the quadratic form as $x_1x_2 + x_3x_4 + \ldots + x_{2l-1}x_{2l}$. Then, putting $L = E_{12} + E_{21} + E_{34} + E_{43} + \ldots$, our Lie algebra consists of the matrices M with $M^\top L + LM = 0$. For $\mathfrak{h}$ we can take the $H = \mathrm{diag}(a_1, -a_1, a_2, -a_2, \ldots, a_l, -a_l)$.

Roots		Root elements
$\omega_i - \omega_j$	$i \neq j$	$E_{2i-1,2j-1} - E_{2j,2i}$
$\omega_i + \omega_j$	$i < j$	$E_{2i-1,2j} - E_{2j-1,2i}$
$-\omega_i - \omega_j$	$i < j$	$E_{2i,2j-1} - E_{2j,2i-1}$

Order in $h_0^\top$ defined by $H_0 = (l-1, l-2, \ldots, 0)$.
Positive roots: The $\omega_i - \omega_j$ and $\omega_i + \omega_j$ with $i < j$.
Fundamental system : $\omega_1 - \omega_2, \omega_2 - \omega_3, \ldots, \omega_{l-1} - \omega_l, \omega_{l-1} + \omega_l$.
Fundamental Weyl chamber: $a_1 > a_2 > \ldots > a_{l-1} > |a_l|$.(Note the absolute value in the last term.)
Maximal root: $\omega_1 + \omega_2, = \omega_1 - \omega_2 + 2(\omega_2 - \omega_3 + \omega_3 - \omega_4 + \ldots + \omega_{l-2} - \omega_{l-1}) + (\omega_{l-1} - \omega_l) + (\omega_{l-1} + \omega_l)$.

The first $l-2$ fundamental roots are connected by links of type A_2 . In addition there is a A_2-link between $\omega_{l-2} - \omega_{l-1}$ and $\omega_{l-1} - \omega_l$, and one between $\omega_{l-2} - \omega_{l-1}$ and $\omega_{l-1} + \omega_l$. Thus the Dynkin diagram is

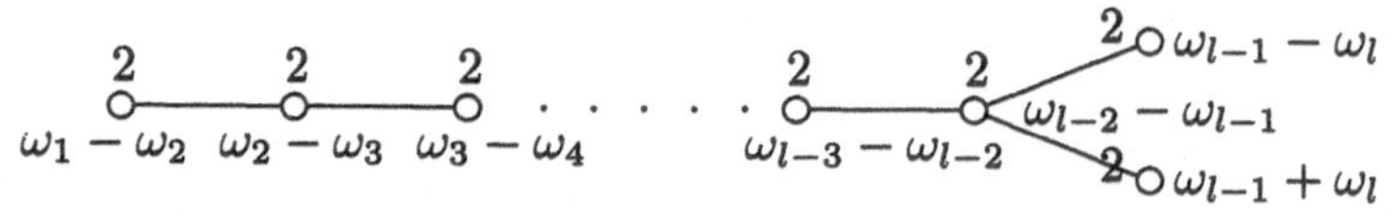

The Killing form is a multiple of $\omega_i{}^2$. The Weyl group contains the interchange of any two axes, corresponding to reflection across $\omega_i - \omega_j = 0$, and also the operation that interchanges two coordinates together with change of their signs, corresponding to reflection in $\omega_i + \omega_j$. Thus it consists of the permutations together with an even number of sign changes of l variables; its order is $2^{l-1} \cdot l!$.

Fundamental coroots: $H_1 = e_1 - e_2, \ldots, H_{l-1} = e_{l-1} - e_l, H_l = e_{l-1} + e_l$.

The Cartan matrix differs from that of A_l by having $a_{l-1,l} = a_{l,l-1} = 0$ and $a_{l-2,l}$ and $a_{l,l-2}$ equal to -1.

(With $\sum_0^{2l} x_i{}^2$ as quadratic form, a Cartan sub Lie algebra is formed by all matrices $\mathrm{diag}(a_1J_1, a_2J_1, \ldots, a_lJ_1)$, the a_i again purely imaginary for $\mathfrak{h}_0$.)

We proceed to describe the root systems, fundamental systems, Dynkin diagrams, and Cartan matrices for the exceptional Lie algebras.

5) G_2.

$\mathfrak{h}$ is the subspace of $\mathbb{C}^3$ with equation $\omega_1 + \omega_2 + \omega_3 = 0$, and $\mathfrak{h}_0$ is the corresponding subspace of $\mathbb{R}^3$. (Vectors in $\mathbb{C}^3$ are written (a_1, a_2, a_3).)
The roots are the restrictions to $\mathfrak{h}$ of $\pm\omega_i$ and $\pm(\omega_i - \omega_j)$. Order in $\mathfrak{h}_0^{\top}$ defined by $(2, 1, -3)$.
Positive roots: $\omega_1, \omega_2, -\omega_3, \omega_1 - \omega_2, \omega_2 - \omega_3, \omega_1 - \omega_3$.
Fundamental system: $\omega_2, \omega_1 - \omega_2$.
Fundamental Weyl chamber: $a_1 > a_2 > 0$.
Maximal root: $\omega_1 - \omega_3, = 3\omega_2 + 2(\omega_1 - \omega_2)$.

ω_2 is a short root, of norm square 1; $\omega_1 - \omega_2$ is a long root, of norm square 3. We can add ω_2 three times to $\omega_1 - \omega_2$ (the arrow goes from $\omega_1 - \omega_2$ to ω_2). The Dynkin diagram is

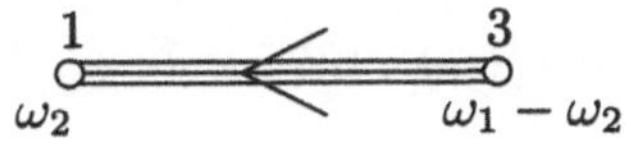

The Killing form is again $k \cdot \sum \omega_i{}^2$. The Weyl group contains the interchange of any two coordinates, corresponding to $\omega_i - \omega_j$ (these act as reflections in $\mathfrak{h}_0$), and so all permutations; it also contains the rotations of $\mathfrak{h}_0$ by multiples of $\pi/3$, in particular the element $-id$.It is isomorphic with the dihedral group $\mathcal{D}_6$. Its order is 12, in agreement with the fact that there are twelve chambers in the $C - S$ diagram. [For the computation we note that the operation associated with ω_3 sends (a_1, a_2, a_3) to $(-a_2, -a_1, -a_3)$.]

Fundamental coroots: $H_1 = (1, -1, 0), H_2 = (-1, 2, -1)$.

The Cartan matrix is

$$\begin{bmatrix} 2 & -1 \\ -3 & 2 \end{bmatrix}$$

Actually all this is part of an explicit description of G_2 as sub Lie algebra of B_3, i.e., $\mathfrak{o}(7, \mathbb{C})$: Let $Y_i, Y_{-i}, Y_{i,-j}, \ldots$ be the root elements of B_3 as in the table for B_l above, and put $Z_{\pm 1} = Y_{\pm 1} \pm Y_{\mp 2, \mp 3}$ etc. (permute cyclically). Then the subspace of B_3 spanned by the $Z_{\pm i}$, the $Y_{j,-k}$ and the subspace $\mathfrak{h}'$ of $\mathfrak{h}$ defined by $\omega_1 + \omega_2 + \omega_3 = 0$ is a sub Lie algebra of B_3, isomorphic to G_2, with $\mathfrak{h}'$ as CSA and the restrictions to $\mathfrak{h}'$ of the $\pm\omega_i$ and the $\omega_j - \omega_k$ as roots. (Note $\omega_1 = -\omega_2 - \omega_3$ etc. on $\mathfrak{h}'$.)

6) F_4.

$\mathfrak{h}$ is $\mathbb{C}^4$, and $\mathfrak{h}_0$ is $\mathbb{R}^4$.

The roots are the forms $\pm\omega_i$ and $\pm\omega_i \pm \omega_j$ with $i, j = 1, 2, 3, 4$ and $i < j$, and the forms $1/2(\pm\omega_1 \pm \omega_2 \pm \omega_3 \pm \omega_4)$. Order in $\mathfrak{h}_0^\top$ defined by $(8, 4, 2, 1)$.
Positive roots: $\omega_i, \omega_i + \omega_j$ and $\omega_i - \omega_j$ with $i < j$, $1/2(\omega_1 \pm \omega_2 \pm \omega_3 \pm \omega_4)$.
Fundamental system: $\alpha_1 = 1/2(\omega_1 - \omega_2 - \omega_3 - \omega_4), \alpha_2 = \omega_4, \alpha = \omega_3 - \omega_4, \alpha_4 = \omega_2 - \omega_3$.
Fundamental Weyl chamber: $a_1 > a_2 + a_3 + a_4, a_4 > 0, a_3 > a_4, a_2 > a_3$.
Maximal root: $\omega_1 = \omega_2, = 2\alpha_1 = 4\alpha_2 + 3\alpha_3 + 2\alpha_4$.

We can add α_2 twice to α_3. The Dynkin diagram is

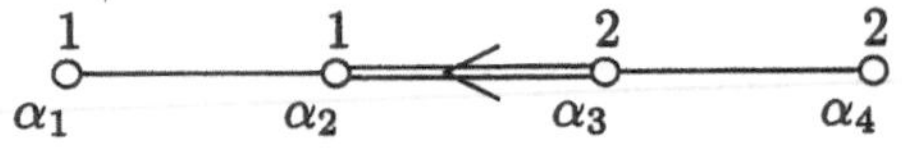

The Killing form is Pythagorean, k $\cdot \sum \omega_i{}^2$. The Weyl group contains all permutations of the axes (from the $\omega_i - \omega_j$), all sign changes (from the ω_i) and the transformation that sends $H = (a_1, a_2, a_3, a_4)$ to $H - (a_1 + a_2 + a_3 + a_4) \cdot E$ with $E = (1, 1, 1, 1)$ (from $1/2(\omega_1 + \omega_2 + \omega_3 + \omega_4)$), and is generated by these elements. Its order is $4! \cdot 2^4 \cdot 3$ (as determined by Cartan [3]).

Fundamental coroots:$H_1 = e_1 - e_2 - e_3 - e_4$, $H_2 = 2e_4, H_3 = e_3 - e_4, H_4 = e_2 - e_3$.

The Cartan matrix differs from that for A_4 only by having -2 as 3, 2-entry.

For E_6, E_7, E_8 we first give Cartan's description. Then follows a more recent model for E_8, in which E_6 and E_7 appear as sub Lie algebras.

7) E_6.

$\mathfrak{h}$ is $\mathbb{C}^6$, and $\mathfrak{h}_0$ is $\mathbb{R}^6$.
The roots are the $\omega_i - \omega_j$, the $\pm(\omega_i + \omega_j + \omega_k)$
with $i < j < k$, and $\pm(\omega_1 + \omega_2 + \cdots + \omega_6)$.
Order in $\mathfrak{h}_0^\top$ defined by $(5, 4, \ldots, 0)$.
Positive roots: $\omega_i - \omega_j$ with $i < j$, $\omega_i + \omega_j + \omega_k$ with $i < j < k$, $\omega_1 + \cdots + \omega_6$.
Fundamental system: $\alpha_1 = \omega_1 - \omega_2, \alpha_2 = \omega_2 - \omega_3, \ldots, \alpha_5 = \omega_5 - \omega_6, \alpha_6 = \omega_4 + \omega_5 + \omega_6$.
Fundamental Weyl chamber: $a_1 > a_2 > \cdots > a_6, a_4 + a_5 + a_6 > 0$.
Maximal root: $\omega_1 + \omega_2 + \cdots + \omega_6, = \alpha_1 + 2\alpha_2 + 3\alpha_3 + 2\alpha_4 + \alpha_5 + 2\alpha_6$.

We can add each α_i to the preceding one once, up to α_5; and we can add α_3 and α_6. The Dynkin diagram is

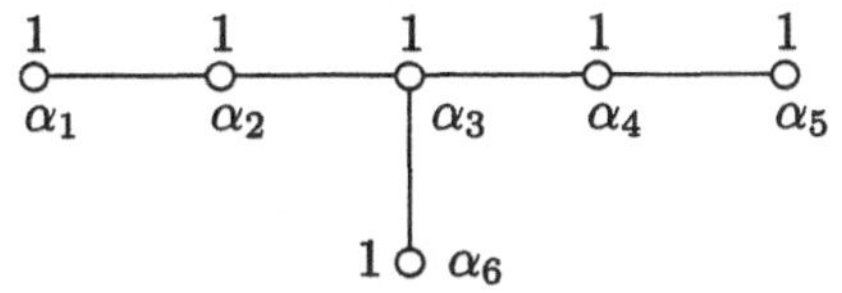

The Killing form is not Pythagorean; it is $24\sum \omega_i{}^2 + 8(\sum \omega_i)^2$.

Order of the Weyl group, as determined by Cartan: $72 \cdot 6!$ (see [3]).

Fundamental coroots: $H_1 = e_1 - e_2, \ldots, H_5 = e_5 - e_6, H_6 = 1/3(-e_1 - e_2 - e_3 + 2e_4 + 2e_5 + 2e_6)$

(One could consider $\mathfrak{h}$ as the subspace $\sqrt{3} \cdot \omega_0 + \omega_1 + \omega_2 + \cdots + \omega_6 = 0$ of $\mathbb{C}^7$ (with coordinates $a_0, a_1, \ldots, a_6$) and Pythagorean metric.)

For the Cartan matrix and another description see below.

8) E_7.

$\mathfrak{h}$ is $\mathbb{C}^7$, and $\mathfrak{h}_0$ is $\mathbb{R}^7$.
The roots are the $\omega_i - \omega_j$, the $\pm(\omega_i + \omega_j + \omega_k)$ with $i < j < k$, and the $\pm \sum_{r \neq i} \omega_r$.
Order in $\mathfrak{h}_0^\top$ defined by $(6, 5, \ldots, 0)$.
Positive roots: $\omega_i - \omega_j$ with $i < j$, $\omega_i + \omega_j + \omega_k$ with $i < j < k$, $\sum_{r \neq i} \omega_r$.
Fundamental system: $\alpha_1 = \omega_1 - \omega_2, \ldots, \alpha_6 = \omega_6 - \omega_7, \alpha_7 = \omega_5 + \omega_6 + \omega_7$.
Fundamental Weyl chamber: $a_1 > a_2 > \cdots > a_7, a_5 + a_6 + a_7 > 0$.
Maximal root: $\omega_1 + \cdots + \omega_6, = \alpha_1 + 2\alpha_2 + 3\alpha_3 + 4\alpha_4 + 3\alpha_5 + 2\alpha_6 + 2\alpha_7$.

We can add each α_i to the preceding one once, up to α_6, and we can add α_4 and α_7. The Dynkin diagram is

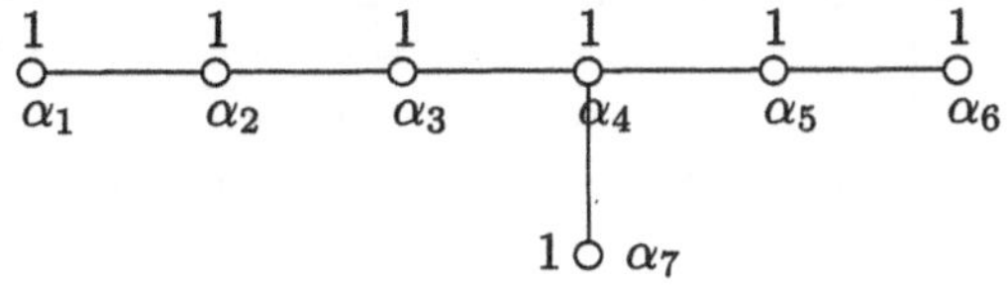

The Killing form is not Pythagorean. (One could consider $\mathfrak{h}$ as the subspace $\sqrt{2} \cdot \omega_0 + \omega_1 + \cdots + \omega_7 = 0$ of $\mathbb{C}^8$ with Pythagorean metric.)

Order of the Weyl group, as determined by Cartan: $56 \cdot 27 \cdot 16 \cdot 10 \cdot 6 \cdot 2$ (see [3]).

Fundamental coroots:$H_1 = e_1 - e_2, \ldots, H_6 = e_6 - e_7, H_7 = 1/3(-e_1 - e_2 - e_3 - e_4 + 2e_5 + 2e_6 + 2e_7)$.

For another description and the Cartan matrix see below.

9) E_8.

For $\mathfrak{h}$ we take the subspace $\omega_1+\omega_2+\cdots+\omega_9=0$ of $\mathbb{C}^9$, with $\mathfrak{h}_0=\mathfrak{h}\cap\mathbb{R}^9$. Roots: the $\omega_i-\omega_j$ with $i\neq j$ and the $\pm(\omega_i+\omega_j+\omega_k)$ with $1\leq i<j<k\leq 9$.
Order in $\mathfrak{h}_0^\top$ defined by $(8,7,\ldots,1,-36)$.
Positive roots: $\omega_i-\omega_j$ with $i<j, \omega_i+\omega_j+\omega_k$ with $i<j<k<9$, and $-\omega_i-\omega_j-\omega_9$ with $i<j<9$.
Fundamental system: $\alpha_1=\omega_1-\omega_2,\ldots,\alpha_7=\omega_7-\omega_8, \alpha_8=\omega_6+\omega_7+\omega_8$.
Maximal root: $\omega_1-\omega_2, \doteq 2\alpha_1+3\alpha_2+4\alpha_3+5\alpha_4+6\alpha_5+4\alpha_6+2\alpha_7+3\alpha_8$.

We can add α_2 to α_1 etc. up to α_7, and we can add α_5 and α_8.
The Dynkin diagram is

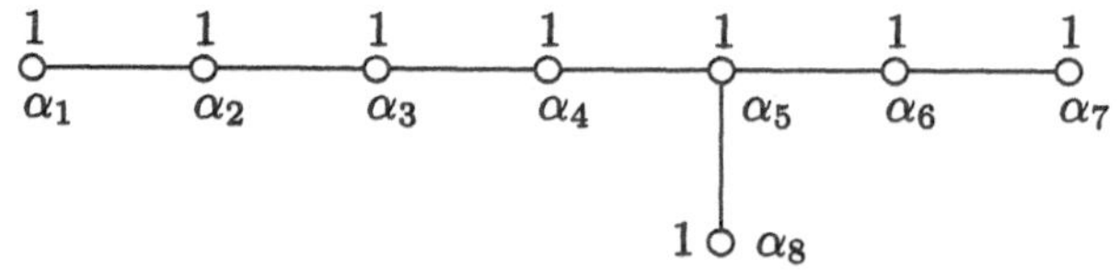

(This diagram appears in many other contexts in mathematics.)

Order of the Weyl group (after Cartan [3]): $240\cdot 56\cdot 27\cdot 16\cdot 10\cdot 6\cdot 2$.

We write out the Cartan matrix (denoted by E_8):

$$E_8=\begin{bmatrix} 2 & -1 & 0 & 0 & 0 & 0 & 0 & 0 \\ -1 & 2 & 0 & & & & & 0 \\ 0 & -1 & 2 & -1 & & & & 0 \\ 0 & 0 & -1 & 2 & -1 & & & 0 \\ 0 & 0 & 0 & -1 & 2 & -1 & 0 & -1 \\ 0 & & & 0 & -1 & 2 & -1 & 0 \\ 0 & & & & 0 & -1 & 2 & 0 \\ 0 & & & 0 & -1 & 0 & 0 & 2 \end{bmatrix}$$

This is an interesting matrix (discovered by Korkin and Zolotarev 1873, [16]). It has integral entries, is symmetric, positive definite (the quadratic form $v^\top E_8 v$, with v in $\mathbb{R}^8$, is positive except for $v=0$), unimodular (i.e. $\det E_8=1$), and of type II or even (the diagonal elements are even; the value $v^\top E_8 v$ with v in $\mathbb{Z}^8$ is always even), and it is the only 8×8 matrix with these properties, up to equivalence (i.e. up to replacing it by $M^\top E_8 M$ with any integral matrix M with $\det M=\pm 1$).

The Cartan matrices for E_7 and E_6 are obtained from that for E_8 by removing the first row and column, resp the first two rows and columns.

Fundamental coroots:$H_1 = e_1 - e_2, \ldots, H_6 = e_6 - e_7, H_7 = e_7 - e_8, H_8 = e_6 + e_7 + e_8 - 1/3e$, with $e = (1, \ldots, 1)$.

There is an alternative description of E_8, with $\mathfrak{h} = \mathbb{C}^8$.
Roots: $\pm\omega_i \pm \omega_j$ with $i \neq j$, and $1/2(\pm\omega_1 \pm \omega_2 \pm \cdots \pm \omega_8)$ with an even number of plus-signs.
Order defined by $H_0 = (0, -1, -2, \ldots, -6, 23)$
Positive roots: $\pm\omega_i - \omega_j$ with $1 \leq i < j < 8$, $\pm\omega_i + \omega_8$ with $1 \leq i < 8$,$1/2(\pm e_1 \pm e_2 \pm \ldots \pm e_7 + e_8)$ (even number of —-signs).
Fundamental system: $\alpha_1 = 1/2 \sum \omega_i, \alpha_2 = -\omega_1 - \omega_2, \alpha_3 = \omega_2 - \omega_3, \alpha_4 = \omega_1 - \omega_2, \alpha_5 = \omega_3 - \omega_4, \alpha_6 = \omega_4 - \omega_5, \alpha_7 = \omega_5 - \omega_6, \alpha_8 = \omega_6 - \omega_7$.
Maximal root: $\omega_8 - \omega_7$

We can add α_4 to α_3.

We show the Dynkin diagram once more, in reversed position and with new numbering:

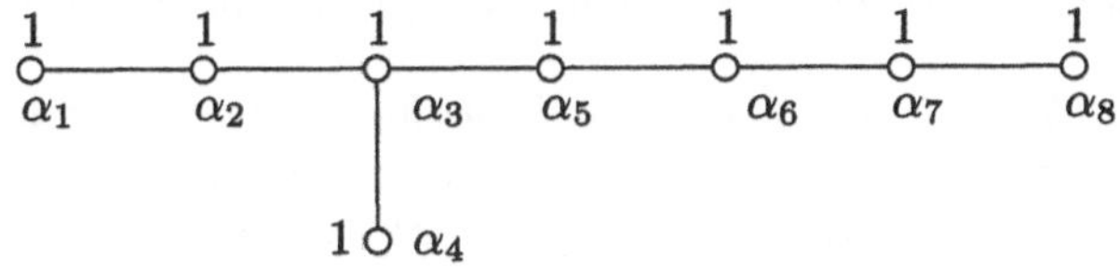

Fundamental coroots:$H_1 = 1/2 \sum e_i, H_2 = -e_1 - e_2, H_3 = e_2 - e_3, H_4 = e_1 - e_2, H_5 = e_3 - e_4, H_6 = e_4 - e_5, H_7 = e_5 - e_6, H_8 = e_6 - e_7$.

The Cartan matrix in this scheme is derived from the earlier one by rearranging rows and columns by the permutation which describes the change, namely $(1, \ldots, 8) \to (8, 7, 6, 5, 3, 2, 1, 4)$.

There are models for E_6 and E_7 in terms of E_8 (we stay with the alternative picture): The root system of E_7 is isomorphic to the subset of the root system of E_8 consisting of those roots that do not involve α_8 when written as linear combinations of the α_i. Similarly the root system of E_6 "consists" of those roots of E_8 that involve neither α_8 nor α_7. More than that, E_7 is (isomorphic to) the sub Lie algebra of E_8 formed by all H_α and $X_{\pm\alpha}$ for the α that do not involve α_8; this sub Lie algebra is generated by the $X_{\pm i}$ with $1 \leq i \leq 7$. Similarly for E_6 one omits α_8 and α_7.

In general, if for a semisimple $\mathfrak{g}$ one takes a subdiagram of the Dynkin diagram obtained by omitting some of the vertices (and the incident edges), then the $X_{\pm i}$ of $\mathfrak{g}$ corresponding to the subdiagram generate a sub Lie algebra of $\mathfrak{g}$ which is semisimple and has precisely the subdiagram as Dynkin diagram.To prove this one should verify that each sub Lie algebra corresponding to one of the components of the subdiagram is simple (the ideal generated by any non-zero element is the whole Lie algebra; use Prop.A(d) of §2.11).

The Cartan matrices for E_7 and E_6 in this system are obtained from that for E_8 by omitting the last row and column or the last two rows and columns.

2.15 Automorphisms

We continue with our semisimple Lie algebra $\mathfrak{g}$ (over $\mathbb{C}$), with a Cartan sub Lie algebra $\mathfrak{h}$, the associated root system Δ, the Weyl group $\mathcal{W}$, etc. The first thing we prove is that the operations of the Weyl group in $\mathfrak{h}$ are induced by inner automorphisms of $\mathfrak{g}$. To recall, an inner automorphism of $\mathfrak{g}$ is a product of a finite number of automorphisms of the form $\exp(\operatorname{ad} X)$ with X in $\mathfrak{g}$. We write $Int(\mathfrak{g})$ for the group formed by all inner automorphisms; this is a subgroup of the group $Aut(\mathfrak{g})$ of all automorphisms of $\mathfrak{g}$ which in turn is a subgroup of the general linear $GL(\mathfrak{g})$ of (the vector space) $\mathfrak{g}$.

THEOREM A. *To any element S of the Weyl group of $\mathfrak{g}$ there exists an inner automorphism A of $\mathfrak{g}$ under which the Cartan sub Lie algebra $\mathfrak{h}$ is stable and for which the restriction $A|\mathfrak{h}$ of A to $\mathfrak{h}$ equals S (as operator on $\mathfrak{h}$).*

For the proof we shall use elementary facts about Lie groups, without much of a definition or proof (see §1.3). The prime example, and the starting point of the proof, is $\mathfrak{sl}(2,\mathbb{C})$, with $\mathfrak{h} = ((H))$ (see §1.1). The Weyl group is $\mathbb{Z}/2$; the non-trivial element T sends H to $-H$.

The Lie group, of which $\mathfrak{sl}(2,\mathbb{C})$ is the Lie algebra, is the special linear group $SL(2,\mathbb{C})$. In it we find the element $J_1 (= X_+ - X_-)$, which conjugates H to $-H$.

Now J_1 can be written as $\exp(\pi/2 \cdot J_1)$, by the familiar computation with series that shows $\exp(it) = \cos t + i \cdot \sin t$, because ${J_1}^2$ is $-I$. This suggests to use as the A of our theorem for the present case the inner automorphism $\exp(t \cdot \operatorname{ad}(X_+ - X_-))$ for a suitable t-value. Indeed, the operator $1/2 \operatorname{ad} J_1$ has matrix $\operatorname{diag}(0, J_1)$ wr to the basis $\{J_1, P, -H\}$ of $\mathfrak{sl}(2,\mathbb{C})$ (here P is $X_+ + X_-$, see p.75). Then we have $\exp(\pi/2 \operatorname{ad} J_1) = \operatorname{diag}(1,-1,-1)$, since $\exp(\pi J_1)$ equals $-I$, and this sends H to $-H$.

We now consider our general $\mathfrak{g}$. Let S_α be the reflection in $\mathfrak{h}$ associated with the root α. Recall the sub Lie algebra $\mathfrak{g}^{(\alpha)} = ((H_\alpha, X_\alpha, X_{-\alpha}))$. Put temporarily $J_\alpha = X_\alpha - X_{-\alpha}$, and form the inner automorphism $A_\alpha = \exp(\pi/2 \operatorname{ad} J_\alpha)$. Our computation for $\mathfrak{sl}(2,\mathbb{C})$ yields $A_\alpha(H_\alpha) = -H_\alpha$. For any H orthogonal to H_α, i.e. for any H with $\alpha(H) = 0$ we have $\operatorname{ad} J_\alpha(H) = 0$ [from $[HX_{\pm\alpha}] = \pm\alpha(H)X_{\pm\alpha}$] and so $A_\alpha(H) = H$. Thus A_α sends $\mathfrak{h}$ to itself and agrees on $\mathfrak{h}$ with S_α. Now the S_α generate the Weyl group, and Theorem A follows.

There is a kind of converse to this.

THEOREM B. *Let A be an inner automorphism of $\mathfrak{g}$ that sends $\mathfrak{h}$ to itself. Then the restriction of A to $\mathfrak{h}$ is equal to an element of the Weyl group.*

Together the two theorems say that the Weyl group of $\mathfrak{h}$ consists of those operators on $\mathfrak{h}$ that come from the *normalizer* $N_{\mathfrak{h}}$of $\mathfrak{h}$ in $Int(\mathfrak{g})$ (the elements that send $\mathfrak{h}$ to itself). In addition we have

THEOREM C. *An automorphism A of $\mathfrak{g}$ that sends $\mathfrak{h}$ to itself and induces the identity map of $\mathfrak{h}$ is of the form* $\exp(\operatorname{ad} H_0)$ *with a suitable element H_0 of $\mathfrak{h}$.*

Thus such an automorphism is automatically in $Int(\mathfrak{g})$, and the *centralizer* $Z_{\mathfrak{h}}$of $\mathfrak{h}$ in $Int(\mathfrak{g})$ (consisting of the elements that leave $\mathfrak{h}$ pointwise fixed) is the set of all $\exp(\operatorname{ad} H)$ for H in $\mathfrak{h}$ (this is a subgroup, since the H's commute and so $\exp(\operatorname{ad}(H + H'))$ equals $\exp(\operatorname{ad} H)\exp(\operatorname{ad} H')$). We see that $Z_{\mathfrak{h}}$is connected.

Altogether we get

THEOREM D. *The assignment $A \to A|\mathfrak{h}$ sets up an isomorphism of the quotient $N_{\mathfrak{h}}/Z_{\mathfrak{h}}$with the Weyl group $\mathcal{W}$. (And $N_{\mathfrak{h}}/Z_{\mathfrak{h}}$is the group of components of $N_{\mathfrak{h}}$)*

We first prove Theorem C (which is easy) and then comment on Theorem B.

Let then A be as in Theorem C. We recall the fundamental roots α_i. The corresponding coroots H_i and the root elements X_i and X_{-i} generate $\mathfrak{g}$, as we know from §2.11. Thus A is determined by its effect on these elements. By hypothesis we have $A(H_i) = H_i$. Therefore each α_i goes to itself (under $A^\top$), and in turn each X_i and each X_{-i} goes to a multiple of itself, with a (non-zero) factor a_i or b_i. The relation $[X_i X_{-i}] = H_i$ and invariance under A requires $b_i = 1/a_i$. Choose t_i so that $a_i = \exp(t_i)$. Since the roots $\{\alpha_i\}$ are a basis for $\mathfrak{h}$, there exists H_0 in $\mathfrak{h}$ with $\alpha_i(H_0) = t_i$. It follows from $\operatorname{ad} H_0(X_{\pm i}) = \pm t_i X_{\pm i}$ that the automorphism $\exp(\operatorname{ad} H_0)$ agrees with A on the H_i and the $X_{\pm i}$; the two are therefore identical.

Theorem B is a good deal harder to prove and in fact goes beyond the scope of these notes. However we briefly indicate the steps. So let A be an inner automorphism of $\mathfrak{g}$ that sends $\mathfrak{h}$ to itself. Applying A to one of the formulae $[HX_\alpha] = \alpha(H)X_\alpha$ that define the roots and root elements, we get $[AH, AX_\alpha] = \alpha(H)AX_\alpha$ or, replacing H by $A^{-1}H$, $[H, AX_\alpha] = A^\vee\alpha(H)AX_\alpha$. Thus $A^\vee\alpha$ is again a root (and AX_α is a corresponding root element). It follows that $A^\vee$ maps $\mathfrak{h}_0^\top$ to itself; and A maps $\mathfrak{h}_0$ to itself (as a real linear transformation) and permutes the coroots H_α (note that A

leaves the Killing form and the induced isomorphism of $\mathfrak{h}_0$ and $\mathfrak{h}_0^\top$ invariant) and by the same token permutes the Weyl chambers. Since the Weyl group is transitive on the chambers, we can, using Theorem A, find an inner automorphism B that induces an element of the Weyl group on $\mathfrak{h}$ and such that the composition $A' = BA$ preserves the fundamental Weyl chamber C. The next step is to show that $A'|\mathfrak{h}$ is in fact the identity. We note first that the linear map $A'|\mathfrak{h}_0$ has a fix vector (eigenvector with eigenvalue 1) H_0 in C, e.g. the sum of the unit vectors, wr to the Killing form, along the edges of C.

One now introduces a compact form $\mathfrak{u}$ of $\mathfrak{g}$, which one can assume to contain $i\mathfrak{h}_0$ (see §2.10). With the scalars restricted to $\mathbb{R}$ one has $\mathfrak{g} = \mathfrak{u} + i\mathfrak{u}$. One shows now (a long story) that the real sub Lie algebra $\mathfrak{u}$ of $\mathfrak{g}$ generates a compact Lie group A in $Int(\mathfrak{g})$, with Lie algebra $\mathfrak{u}$, and that every element of $Int(\mathfrak{g})$ is (uniquely) of the form $k \cdot \exp(\operatorname{ad} iY)$ with k in K and Y in $\mathfrak{u}$ (analogous to writing any invertible complex matrix as unitary times positive definite Hermitean – the polar decomposition). In particular the automorphism A' above can be so written. Now comes a lemma, which allows one to disregard the iY-term. Note that the fix vector H_0 of A' lies in $\mathfrak{h}_0$ and so in $i\mathfrak{u}$.

LEMMA E. *Suppose for some H in $\mathfrak{h}_0$ the element $k \cdot \exp(\operatorname{ad} iY)(H) = H'$ is also in $\mathfrak{h}_0$. Then $[YH] = 0$, and $\exp(\operatorname{ad} iY)(H) = H$.*

Proof: Since $\mathfrak{u}$ is a real form of $\mathfrak{g}$, complex conjugation of $\mathfrak{g}$ wr to $\mathfrak{u}$ (sending i to $-i$ in the decomposition $\mathfrak{g} = \mathfrak{u} + i\mathfrak{u}$) preserves brackets, and so one has $k \cdot \exp(-\operatorname{ad} iY)(-H) = -H'$. This implies $\exp(\operatorname{ad} 2iY)(H) = H$. Now $\operatorname{ad} Y$ is a skew-symmetric (wr to the Killing form) on $\mathfrak{u}$, and so its eigenvalues on $\mathfrak{u}$ and then also on $\mathfrak{g}$ are purely imaginary. The eigenvalues of $\operatorname{ad} iY$ are then real; it is also semisimple, just as $\operatorname{ad} Y$ is. But then it is clear from the diagonal form of $\operatorname{ad} iY$ that the fix vector H of $\exp(\operatorname{ad} 2iY)$ must be an eigenvector of $\operatorname{ad} iY$ with eigenvalue 0, i.e., must satisfy $\operatorname{ad} Y(H) = 0$, or $[YH] = 0$.

This in turn implies $\exp(\operatorname{ad} sY)(H) = H$ for all s.

Applied to the $A' = k \cdot \exp(\operatorname{ad} iY)$ above this has the consequence $A'(H) = k(H)$ for all H in $\mathfrak{h}_0$, and in particular $k(H_0) = H_0$. Now one has another important fact which we don't prove here. (Cf. [12], Cor. 2.8, p.287.)

PROPOSITION F. *In a compact connected Lie group the stabilizer of any element of the Lie algebra is connected.*
(The *stabilizer* of X is the set (group) $\{g : \operatorname{Ad} g(X) = X\}$. Here $\operatorname{Ad} g$ refers to the *adjoint* action of g on $\mathfrak{g}$, induced by conjugation of G by g, see [11].)

One applies this to the element H_0. Then the elements $\exp(itH_0)$, for real t, which lie in K, commute with k. The fact that no root vanishes on H_0 (or iH_0) implies that the Lie algebra of the stabilizer of H_0 (in $\mathfrak{u}$) is

$i\mathfrak{h}_0$. Thus k lies in $\exp(i\mathfrak{h}_0)$ (by Prop.F), and therefore it and then also A' acts as id on $i\mathfrak{h}_0$ and on $\mathfrak{h}$.

Finally then $A|\mathfrak{h}_0$ equals $B^{-1}|\mathfrak{h}_0$ and is therefore equal to an operator in the Weyl group, establishing Theorem B. √

We now want to go from $Int(\mathfrak{g})$ to $Aut(\mathfrak{g})$. The important concept here is that of a *diagram automorphism.* We recall the basic Isomorphism Theorem (§2.9, Cor.2). It suggests looking at the weak equivalences of the root system Δ of $\mathfrak{g}$ with itself; as noted loc cit, each such equivalence extends uniquely to an isometry of $\mathfrak{h}_0$ with itself, and we will use both aspects interchangeably. Under composition the self-equivalences form a group, a subgroup of the group of all permutations of Δ, called the *automorphism group* of Δ and denoted by $Aut(\Delta)$. It has the Weyl group as a subgroup, in fact as a normal subgroup (the conjugate of a Weyl reflection S_α by an element T in $Aut(\Delta)$ is the reflection wr to the root $T(\alpha)$). There is also the subgroup of those elements that send the fundamental Weyl chamber to it self, or—equivalently—permute the fundamental roots among themselves; it can also be interpreted as the group of automorphisms (in the obvious sense) of the Dynkin diagram; we denote it by $Aut(DD)$. (See §2.13.)

Since $\mathcal{W}$ is simply transitive on the chambers, it is clear that $Aut(\Delta)$ is the semidirect product of $\mathcal{W}$ and $Aut(DD)$, and that $Aut(DD)$ can be identified with the quotient group $Aut(\Delta)/\mathcal{W}$.

The basic isomorphism theorem cited above allows us to associate with each element of $Aut(\Delta)$ an automorphism of $\mathfrak{g}$. However there are choices involved, and one does not get a group of automorphisms of $\mathfrak{g}$ this way. This is different if one restricts oneself to $Aut(DD)$. An element T of it permutes the fundamental roots α_i in a certain way; one gets an associated automorphism A_T of $\mathfrak{g}$ by permuting the corresponding root elements X_i and X_{-i} (which generate $\mathfrak{g}$) in the same way. It is now clear that the map $T \twoheadrightarrow A_T$ is multiplicative. The automorphisms of $\mathfrak{g}$ so obtained from $Aut(DD)$ are called *diagram automorphisms.*

This depends of course on the choice of $\mathfrak{h}$ and of the fundamental Weyl chamber. However, for any two fundamental systems Φ and Φ' we know that there exist inner automorphisms that send Φ to Φ' and that the map $\Phi \to \Phi'$ so obtained is unique (Propositions C, D, E, §2.11, and Theorems A, B and C); thus we can identify all fundamental systems of $\mathfrak{g}$ to a *generic* fundamental system, with a corresponding generic Dynkin diagram. It is easily seen that any automorphism of $\mathfrak{g}$ induces a well-defined automorphism of the generic fundamental system and Dynkin diagram, and that this yields a homomorphism of $Aut(\mathfrak{g})$ into $Aut(DD)$ (the latter now interpreted as the group of automorphisms of the generic Dynkin diagram). Theorem C implies that the kernel of this map is precisely $Int(\mathfrak{g})$. The diagram automorphisms above show that $Aut(\mathfrak{g})$ contains a subgroup that maps isomorphically onto $Aut(DD)$. We now have a good hold on the relation between $Aut(\mathfrak{g})$ and $Int(\mathfrak{g})$:

THEOREM G. *The sequence* $1 \to Int(\mathfrak{g}) \to Aut(\mathfrak{g}) \to Aut(DD) \to 1$ *is split exact.*

Another way to put (part of) this is to say that "the group $Out(\mathfrak{g})$ of outer automorphisms of $\mathfrak{g}$ ", i.e., the quotient group $Aut(\mathfrak{g})/Int(\mathfrak{g})$, can be identified with $Aut(DD)$.

We noted already in effect in §2.13 what $Aut(DD)$ is for the Dynkin diagrams of the various simple Lie algebras: $A_1, B_l, C_l, G_2, F_4, E_7, E_8$ admit only the identity (and so all automorphisms of the Lie algebra are inner). A_l for $l > 1$, D_l for $l \neq 4$ and E_6 admit one other automorphism ("horizontal" reversal for A_l and E_6, interchanging the two "ends" for D_l), so that $Aut(DD)$ is $\mathbb{Z}/2$; and finally D_4 permits the full symmetric group $\mathcal{S}_3$ on three objects (the endpoints of its diagram). (The non-trivial element of $Aut(DD)$ is induced for $\mathfrak{sl}(n, \mathbb{C}), n > 2$, by the automorphism "infinitesimal contragredience", $X \to X^{\triangle}$, and for $\mathfrak{o}(2n, \mathbb{C})$ by conjugation with the improper orthogonal matrix diag$(1, \ldots, 1, -1)$; as noted, $\mathfrak{o}(8, \mathbb{C})$ has some other outer automorphisms in addition.)

Our final topic is the so-called *opposition element* in the Weyl group of any $\mathfrak{g}$. It is that element of the Weyl group that sends the fundamental Weyl chamber C to its negative, $-C$; we denote this element by *op*. Clearly, if $\mathcal{W}$ contains the element $-id$, then *op* is $-id$. This is necessarily so for a $\mathfrak{g}$ with trivial $Aut(DD)$: For $\mathfrak{g}$ we have the contragredience automorphism $C^{\vee}$ of §2.9 (end), whose restriction to $\mathfrak{h}$ is $-id$. By the results above, $C^{\vee}$ is inner iff $-id$ is in the Weyl group; and for a $\mathfrak{g}$ with trivial $Aut(DD)$ all automorphisms are inner.

For D_l with even l the element $-id$ is in $\mathcal{W}$.

For A_l, with $l > 1$, where $\mathcal{W}$ acts as the symmetric group on the coordinate functions ω_i, *op* is the permutation $\omega_i \to \omega_{l+2-i}$. This sends each fundamental root $\alpha_i = \omega_i - \omega_{i+1}$ to $\omega_{l+2-i} - \omega_{l+1-i} = -\alpha_{l+1-i}$, and so sends the fundamental chamber to its negative.

For D_l with odd l the opposition is given by $\omega_i \to -\omega_i$ for $i = 1, \ldots, l-1$ and $\omega_l \to \omega_l$. This sends the fundamental roots $\alpha_i = \omega_i - \omega_{i+1}$ with $i = 1, \ldots, l-2$ to their negatives, and sends $\alpha_{l-1} = \omega_{l-1} - \omega_l$ [resp $\alpha_l = \omega_{l-1} + \omega_l$] to $-\alpha_l$ [resp $-\alpha_{l-1}$], thus sending C to $-C$.

We come to E_6. First a general fact: for any root α and any weight λ the element $S_\alpha(\lambda) - \lambda = \lambda(H_\alpha)\alpha$ lies in the root lattice $\mathcal{R}$ (see §3.1); it follows, using the invariance of $\mathcal{R}$ under $\mathcal{W}$, that $S(\lambda) - \lambda$ lies in $\mathcal{R}$ for any S in $\mathcal{W}$. We use this to show that $-id$ is not in the Weyl group of E_6; namely, for the fundamental weight λ_1 (see §3.5)the element $-id(\lambda_1) - \lambda_1 = -2\lambda_1 = -2/3 \cdot (4\omega_1 + \omega_2 + \cdots + \omega_6)$ is not in $\mathcal{R}$.

In all three cases we have $op \neq -id$, $-id$ is not in the Weyl group, $-op$ gives a non trivial element of $Aut(DD)$, and $C^{\vee}$ is not an inner automorphism.

3

Representations

This chapter brings the construction of the finite dimensional representations of a complex semisimple Lie algebra from the root system. (The main original contributors are É. Cartan [3], H. Weyl [25,26], C. Chevalley [5], Harish-Chandra [10].) We list the irreducible representations for the simple Lie algebras. Then follows Weyl's character formula, and its consequences (the dimension formula, multiplicities of weights of a representation and multiplicities of representations in a tensor product). A final section determines which representations consist of orthogonal, resp symplectic, matrices (in a suitable coordinate system).

Throughout the chapter $\mathfrak{g}$ is a complex semisimple Lie algebra of rank l, $\mathfrak{h}$ is a Cartan sub Lie algebra, $\Delta = \{\alpha, \beta, \ldots\}$ is the root system and Δ^+ is the set of positive roots wr to some given weak order in $\mathfrak{h}$, $\Phi = \{\alpha_1, ..., \alpha_l\}$ is the fundamental system, H_α (with α in Δ) are the coroots, h_α are the root vectors, X_α are the root elements, and the coefficients $N_{\alpha\beta}$ are in normal form (all as described in Ch.2). As noted in §2.11, we write H_i instead of H_{α_i}, for α_i in Φ, for the fundamental coroots; Θ denotes the set $\{H_1, H_2, \ldots, H_l\}$. Similarly we write X_i for X_{α_i} and X_{-i} for $X_{-\alpha_i}$.

3.1 The Cartan-Stiefel diagram

This is a preliminary section, which extends the considerations of §2.11 and introduces some general definitions and facts. For all of it we could replace Δ (in $\mathfrak{h}_0^\top$) by an abstract root system (in a Euclidean space V), with $\mathfrak{h}_0$ corresponding to the dual space $V^\top$ (using the standard identification of a vector space with its second dual). In the literature $\mathfrak{h}_0^\top$ and $\mathfrak{h}_0$ are often *identified* under the correspondence $\lambda \leftrightarrow h_\lambda$ given by the metric; but we shall keep them separate.

We recall that $\{h_\alpha\}$ and Δ are congruent root systems and that $\{H_\alpha\}$ is the root system dual to $\{h_\alpha\}$. $\{H_\alpha\}$ and $\{h_\alpha\}$ have the same Weyl group, isomorphic to that of Δ in the obvious way (contragredience; the reflection for H_α equals that for h_α).

We note that Θ is a fundamental system for the root system $\{H_\alpha\}$: Each relation $\alpha = \sum a_i\alpha_i$ for α in Δ^+, with non-negative integral a_i, implies the relation $\langle\alpha,\alpha\rangle H_\alpha = \sum a_i\langle\alpha_i,\alpha_i\rangle H_i$ (because of $\langle\alpha,\alpha\rangle H_\alpha = 2h_\alpha$ etc.). Thus all these H_α lie in the cone spanned by Θ, and that is of course enough to establish our claim; it also follows that the numbers $a_i\langle\alpha_i,\alpha_i\rangle/\langle\alpha,\alpha\rangle$ are (non-negative) integers.

We regard the Weyl group W as an abstract group, associated to $\mathfrak{g}$, which acts on $\mathfrak{h}_0^\top$ (with the original definition of §2.6) and also on $\mathfrak{h}_0$ with the contragredient (transposed-inverse) action. Thus we have $S\lambda(H) = \lambda(S^{-1}H)$ for S in $\mathcal{W}$, λ in $\mathfrak{h}_0^\top$, and H in $\mathfrak{h}_0$. Since the inner products in $\mathfrak{h}_0$ and $\mathfrak{h}_0^\top$ are compatible ($|\alpha| = |h_\alpha|$), the action on $\mathfrak{h}_0$ is also orthogonal, and in particular each S_α acts as reflection across the *singular plane* $(\alpha, 0) = \{H : \alpha(H) = 0\}$ (cf.§2.11). The formula for this is $S_\alpha(H) = H - \alpha(H)H_\alpha$.

We recall that the union over Δ of these singular planes is the infinitesimal Cartan-Stiefel diagram D' of $\mathfrak{g}$ (in $\mathfrak{h}_0$). It divides $\mathfrak{h}_0$ into the Weyl chambers. The fundamental Weyl chamber C consists of all the H in $\mathfrak{h}_0$ for which the values $\alpha_i(H)$, or equivalently the $\langle H_i, H\rangle$, or again all the $\alpha(H)$ with α in Δ^+, are positive. Similarly the fundamental Weyl chamber $C^\top$ in $\mathfrak{h}_0^\top$ consists of the λ with all $\lambda(H_i)$ positive. The Weyl chambers are cones, of the linear kind described in the Appendix. The walls of the fundamental chamber lie in the planes orthogonal to the H_i (in $\mathfrak{h}_0$), resp. the α_i (in $\mathfrak{h}_0^\top$). (As examples see the figures for the cases A_2, B_2, G_2 in §3.5.)

We come to the new definitions:

Generalizing the notion of singular plane $(\alpha, 0)$, we define, for α in Δ and n in $\mathbb{Z}$, the *singular plane* (α, n), of *height* n, as $\{H \in \mathfrak{h}_0 \colon \alpha(H) = n\}$; note $(\alpha, n) = (-\alpha, -n)$. The union over α and n of the (α, n) is the *(global) Cartan- Stiefel diagram* $D(\mathfrak{g})$, or D in short, of $\mathfrak{g}$ (wr to $\mathfrak{h}$; by conjugacy of the CSA's it is independent of which $\mathfrak{h}$ we use). The components of the complement of $D(\mathfrak{g})$ in $\mathfrak{h}_0$ are the *cells* of the diagram.

(We recall that a *lattice* in a vector space is a subgroup (under addition) generated by some basis of the space.) The subgroup of $\mathfrak{h}_0$ generated by all the H_α (or equivalently by Θ) is called the *translation lattice* $\mathcal{T}$. The subgroup of $\mathfrak{h}_0$ of those H for which all values $\alpha(H)$, with α running over Δ (or ϕ), are integers is called the *center lattice* $\mathcal{Z}$. Dually we write $\mathcal{R}$ (the *root lattice*) for the subgroup of $\mathfrak{h}_0^\top$ generated by Δ (or Φ), and $\mathcal{I}$ (the lattice of *integral forms* or *weights*) for the subgroup of $\mathfrak{h}_0^\top$ consisting of the λ for which all values $\lambda(H\alpha)$ with α in Δ (or in Φ, i.e., using only the H_i in Θ) are integers. For examples see §3.6.

Each element t of $\mathcal{T}$ defines a map of $\mathfrak{h}_0$ to itself, called a *translation*, with $H \to H + t$. The group of maps of $\mathfrak{h}_0$ to itself generated by all these translations and by the Weyl group $\mathcal{W}$ is called the *affine* or *extended* Weyl group $\mathcal{W}_a$, with a split exact sequence $0 \to \mathcal{T} \to \mathcal{W}_a \to \mathcal{W} \to 0$. All its elements are isometries - maps that leave the distance between any two points invariant; but they don't necessarily fix the origin (they are *affine* transformations). Clearly each element of $\mathcal{W}_a$ maps the Cartan-Stiefel diagram $D(\mathfrak{g})$ to itself, and thus permutes the cells.

In $\mathcal{I}$ we distinguish two important subsets: First $\mathcal{I}^d$, the set of the λ in $\mathcal{I}$ with all $\lambda(H_i) \geq 0$ (equivalently: $\lambda(H_\alpha) \geq 0$ for all α in Δ^+), the *dominant* forms or weights; second the set $\mathcal{I}^0$ of the l in $\mathcal{I}$ with all $\lambda(H_i) > 0$, the *strongly dominant* forms or weights. One sees that $\mathcal{I}^0$ (resp. $\mathcal{I}^d$) is the intersection of $\mathcal{I}$ with the fundamental Weyl chamber (resp. the closed fundamental Weyl chamber) of Δ in $\mathfrak{h}_0^\top$. We introduce the set $\Lambda = \{\lambda_1, \lambda_2, \ldots, \lambda_l)$ of independent generators of $\mathcal{I}$, namely the dual basis to Θ, defined by the equations $\lambda_i(H_j) = \delta_{ij}$ (Kronecker δ). The λ_i are the *fundamental weights*; they lie on the edges (the 1-dimensional faces) of the fundamental Weyl chamber, since we have $\lambda_i(H_j) = 0$ for $i \neq j$. [λ_i is the point of intersection of the i-th edge with the plane through the point $1/2\alpha_i$, orthogonal to the vector α_i (the factor $1/2$ comes from $\alpha_i(H_i) = 2$).] $\mathcal{I}^0$ (resp. $\mathcal{I}^d$) is the set of linear combinations of the λ_i with positive (resp. non-negative) integral coefficients. $\mathcal{I}^d$ is a free Abelian semigroup, with basis Λ.

We single out an important element of $\mathcal{I}^d$, the element $\delta = \lambda_1 + \lambda_2 + \ldots + \lambda_l$, the *lowest strongly dominant form*, usually just called the *lowest form* (or *lowest weight*; in the literature also often denoted by ρ); it is characterized by the equations $\delta(H_i) = 1$ for $i = 1, \ldots, l$. Clearly a dominant form λ is strongly dominant iff $\lambda - \delta$ is dominant.

We now prove a number of facts about all these objects.

PROPOSITION A. *The fundamental Weyl chamber C (in $\mathfrak{h}_0$) is contained in the cone spanned by the set Θ.*

(Geometrically, because of $\langle H_i, H_j \rangle \leq 0$ for $i \neq j$ the set Θ spans a "wide" cone, and therefore C, the negative of the "dual" cone, is contained in it.) Take $v = \sum r_i H_i$ in C, i.e., with all $\langle v, H_i \rangle \geq 0$. Write v as $v^- + v^+$, where v^- means the sum of the terms with $r_i < 0$. For any H_i that occurs in the sum v^- (and so not in v^+) we have $\langle v^+, H_i \rangle \leq 0$ because of $\langle H_i, H_j \rangle \leq 0$ for $i \neq j$, and so $\langle v^-, H_i \rangle = \langle v, H_i \rangle - \langle v^+, H_i \rangle \geq 0$. Multiplying by the (non-positive) r_i and adding we get $\langle v^-, v^- \rangle \leq 0$, i.e., $v^- = 0$. $\surd$

It follows from the corresponding fact for $\mathfrak{h}_0^\top$ that the fundamental weights λ_i are positive (in the given weak order) and that the lowest form δ is indeed the smallest element of $\mathcal{I}^0$.

PROPOSITION B. *The lowest form δ equals one half the sum of all positive roots. For any S in $\mathcal{W}$ the element $\delta - S\delta$ is the sum of those positive roots that become negative under S^{-1}.*

For the proof we write temporarily $\epsilon = 1/2 \sum_{\Delta^+} \alpha$. By Lemma F, §2.11, we have $S_i(\epsilon) = \epsilon - \alpha_i$ for the Weyl reflection associated to the fundamental root α_i. Comparing with the general formula $S_i(\lambda) = \lambda - \lambda(H_i)\alpha_i$ we find

$\epsilon(H_i) = 1$ for all i; but then ϵ is δ. The second assertion of Prop.B is then elementary. $\surd$

Let $U = \{z : |z| = 1\}$ be the unit circle in $\mathbb{C}$, as multiplicative group (this is just the unitary group $U(1)$). $\mathfrak{h}_0^\top$ and $\mathfrak{h}_0$ are paired to U by the "bilinear" function that sends the pair (λ, H) to $\exp(2\pi i \lambda(H))$. The *annuller* (or *annihilator*) of a subgroup of $\mathfrak{h}_0$ (resp. $\mathfrak{h}_0^\top$) is the subgroup of $\mathfrak{h}_0^\top$ (resp. $\mathfrak{h}_0$) of those elements that under the pairing to U yield 1 for every element in the given subgroup. We use some simple notions of Pontryagin duality theory of Abelian groups: The dual A^* of a (topological) Abelian group A is the group $Hom(A, U)$ of all continuous homomorphisms of A into U (the characters of A), with the pointwise product, with a suitable topology, and the pairing $(f, a) \to f(a)$ to U.

PROPOSITION C. *The groups $\mathcal{T}$, $\mathcal{Z}$, $\mathcal{R}$, $\mathcal{I}$ are lattices (in $\mathfrak{h}_0$ and $\mathfrak{h}_0^\top$ respectively). $\mathcal{T}$ is a subgroup of $\mathbb{Z}$, and $\mathcal{R}$ one of $\mathcal{I}$. $\mathcal{T}$ and $\mathcal{I}$ are annullers of each other, similarly for $\mathcal{R}$ and $\mathcal{Z}$. The groups $\mathcal{Z}/\mathcal{T}$ and $\mathcal{I}/\mathcal{R}$ are finite, and are dual under the induced pairing to U (and thus isomorphic) .*

That $\mathcal{T}$ and $\mathcal{R}$ are lattices, generated by Θ and Φ respectively, we have seen already. $\mathcal{I}$ and $\mathcal{Z}$ are then generated by the dual bases, Λ and an unnamed one for $\mathcal{Z}$. The inclusion relations come from the integrality of the $\beta(H_\alpha)$. The finiteness of the quotients comes from the fact that all four groups have the same rank. That $\mathcal{Z}/\mathcal{T}$ and $\mathcal{I}/\mathcal{R}$ are dual (each "is" the group of all homomorphisms of the other into U), follows easily from the facts claimed about annulling - which are also quite clear ($\mathcal{I}$ is defined as annuller of $\mathcal{T}$; that conversely $\mathcal{T}$ is annuller of $\mathcal{I}$ one can see by using the symmetry in the definition of dual bases). That duality implies (non-natural) isomorphism for finite Abelian groups is well known; it follows from the facts that duality preserves direct sums and that the dual of the finite cyclic group $\mathbb{Z}/n$ is isomorphic to $\mathbb{Z}/n$. $\surd$

We note, but shall not prove, the fact that $\mathcal{Z}/\mathcal{T}$ is (isomorphic to) the center of the simply connected (compact) Lie group with Lie algebra $\mathfrak{u}$ (compact form of $\mathfrak{g}$, §2.10).

The coroots H_α, for α in Δ, are all primitive elements of $\mathcal{T}$ (they are not divisible, in $\mathcal{T}$, by any integer different from ± 1). The reason is that each H_α belongs to some fundamental system and thus to a basis for $\mathcal{T}$ (§2. 11); similarly for the α and $\mathcal{R}$.

The affine Weyl group $\mathcal{W}_a$ contains the reflections in the singular planes (α, n); e.g., the composition of S_α with translation by H_α is the reflection in the plane $(\alpha, 1)$; the "1" comes from $\alpha(H_\alpha) = 2$. It is easily seen that $\mathcal{W}_a$ is in fact generated by these reflections. It follows, as for the chambers under $\mathcal{W}$, that $\mathcal{W}_a$ is transitive over the cells, and that therefore all cells are congruent. Cells are clearly bounded convex sets. The cell in the fundamental Weyl chamber whose closure contains the origin is called the *fundamental cell*, c.

PROPOSITION D. *If $\mathfrak{g}$ is simple, then the fundamental cell is the simplex $\{H : \alpha_i(H) > 0$ for $i = 1, ..., l$ and $\mu(H) < 1\}$, cut off from the fundamental Weyl chamber by the maximal root μ.*

This follows from Prop. L, §2,11.

LEMMA E. *Let t be a non-zero element of $\mathcal{T}$; then there exists a root α with $\alpha(t) \geq 2$.*

We use the notion of *level*: writing an element s of $\mathcal{T}$ as $\sum s_i H_i$ (with s_i in $\mathbb{Z}$), this is $\sum s_i$, the sum of the coefficients. Now suppose all the values $\alpha(t)$ are ± 1 or 0. The same holds then for all transforms St with S in $\mathcal{W}$; thus we may assume the t_i in $t = \sum t_i H_i$ to be non-negative (transform into C and apply Prop.A). From $\langle t, t \rangle = \sum t_i \langle t, H_i \rangle$ we conclude that there is at least one j with $t_j > 0$ and $\langle t, H_j \rangle > 0$; the latter implies $\alpha_j(t) = 1$ by our assumption on $\mathcal{T}$. The element $S_j t = t - \alpha_j(t) H_j = t - H_j$ still has all coefficients non-negative, when written in terms of the H_i. But the level has gone down by 1. Iterating this we end up with a contradiction when we get to a single H_i, since $\alpha_i(H_i) = 2$. $\checkmark$

We now prove, among other things, that $\mathcal{W}_a$ is simply transitive on the set of cells.

PROPOSITION F.

(a) The only element of the affine Weyl group that keeps any cell fixed (setwise) is the identity.

(b) Each closed cell has exactly one point (a vertex) in the lattice $\mathcal{T}$.

(c) The union of the closed cells that contain the origin is a fundamental domain for $\mathcal{T}$.

(d) The only reflections contained in $\mathcal{W}_a$ are those across the singular planes (α, n).

Keeping a cell fixed, in (a), is of course equivalent to the existence of a fixed point in the (open) cell. For the proof we may as well assume that $\mathfrak{g}$ is simple. In the general case the various simple components operate in pairwise orthogonal invariant subspaces and are independent of each other.

First (a): By transitivity we may assume that the cell in question is the fundamental cell c. Suppose that for a T in $\mathcal{W}_a$ we have $T(c) = c$. If T leaves the origin fixed, it leaves the Weyl chamber C fixed (setwise), and by Prop. E, §2.11 we have $T = id$. If $T(0)$ were not 0, it would be an element of $\mathcal{T}$, in C, on which the maximal root μ takes value 1 (by Prop.D), contradicting Lemma E (note that $\alpha(T(0))$ is a non-negative integer for every positive root α).

For (b) suppose that c had another vertex t, besides 0, in $\mathcal{T}$. Translation by $-t$ sends c into another cell c' that also has 0 as a vertex. There exists then an S in $\mathcal{W}$ with $S(c') = c$. By (a) this would say that S equals the translation by t, which is manifestly not so.

Now for (c): Let $Q(= \mathcal{W} \cdot \bar{c})$ denote the set described in (c). Since the closed cells cover $\mathfrak{h}_0$, it follows from (b) that each point of $\mathfrak{h}_0$ can be translated into Q by a suitable element of $\mathcal{T}$. On the other hand, Prop.D implies that for any point H in Q and any root α we have $|\alpha(H)| \leq 1$; i.e., Q is contained in the strip $\{H : |\alpha(H)| \leq 1\}$. Suppose now that H and H' are two points in Q that are equivalent under $\mathcal{T}$, so that $H - H' = t(\neq 0)$ is in $\mathcal{T}$. Lemma E provides a root α with $\alpha(t) \geq 2$. But then we must have $\alpha(H) = -\alpha(H') = 1$, so that both H and H' lie on the boundary of the strip associated with α, and so also on the boundary of Q. Thus Q has the properties required of a fundamental domain for $\mathcal{T}$.

Finally (d) is immediate from (a). $\checkmark$

Remark to (c): One sees easily that the set Q is the intersection, over Δ, of all the strips described. But for some α the strip may contain Q in its interior (e.g. for the short roots of G_2), and for some α the intersection of Q with the boundary of the strip may be (non-empty and) of dimension less than $l - 1$ (e.g., for the short roots of B_2). This corresponds to the fact that in general the roots that occur as maximal roots wr to some weak order form a proper subset of Δ.

3.2 Weights and weight vectors

We now come to the study of representations. (We shall often abbreviate "representation" to "rep" and similarly "irreducible rep" to "irrep".) Let $\varphi : \mathfrak{g} \to \mathfrak{gl}(V)$ be a representation of $\mathfrak{g}$ on the (complex) vector space V. (We often write Xv or $X \cdot v$ for $\varphi(X)(v)$.)

The basic notion is that of *weight vector*: a joint eigenvector of all the operators $\varphi(H)$ for H in the Cartan sub Lie algebra $\mathfrak{h}$. Note that by definition such a vector is not 0. If v is a weight vector, then the corresponding eigenvalue for $\varphi(H)$, as function of H, is a linear function on $\mathfrak{h}$, in other words an element of $\mathfrak{h}^\top$; this element is the *weight* of v.

For a given λ in $\mathfrak{h}^\top$ the *weight space* V_λ is the subspace of V (possibly 0) consisting of 0 and all the weight vectors with λ as weight. λ is called a *weight of* φ if V_λ is not 0, i.e., if there exists a weight vector to λ. The dimension m_λ of V_λ is called the *multiplicity* of λ (as weight of the rep φ).

We prove a simple, but fundamental, lemma (generalizing Lemma A in §1.11, for A_1). Let v be a weight vector of φ, with weight λ; let α be any root, and let X_α be the corresponding root element (well determined up to a scalar factor, see §2.5).

LEMMA A. *The vector $X_\alpha v$, if not zero, is again a weight vector of φ, with weight $\lambda+\alpha$; in other words, X_α maps V_λ into $V_{\lambda+\alpha}$.*

This is a trivial computation; again, as the physicists say, we "use the commutation rules": From $\varphi([HX_\alpha]) = \varphi(H)\varphi(X_\alpha) - \varphi(X_\alpha)\varphi(H)$ (since φ preserves brackets) and $[HX_\alpha] = \alpha(H)X_\alpha$ (since X_α is root element to α) we get $HX_\alpha v = X_\alpha Hv + [HX_\alpha]v = X_\alpha\lambda(H)v + \alpha(H)X_\alpha v = (\lambda(H) + \alpha(H)) \cdot X_\alpha$. $\surd$

We come to the basic facts about weights, with φ and V as above.

THEOREM B.

(a) V is spanned by weight vectors; there is only a finite number of weights;

(b) the weights are integral forms (they belong to the lattice $\mathcal{I}$ in $\mathfrak{h}_0^\top$);

(c) the set of weights of φ is invariant under the Weyl group: if φ is a weight, so is $S_\alpha\lambda = \lambda - \lambda(H_\alpha)\alpha$, for any α in Δ; in fact, with $\epsilon =$ sgn$(\lambda(H_\alpha))$, all the terms $\lambda, \lambda-\epsilon\alpha, \lambda-2\epsilon\alpha, ..., \lambda-\lambda(H_\alpha)\alpha$ are weights of φ;

(d) the multiplicities are invariant under the Weyl group: $m_\lambda = m_{S\lambda}$ for all S in $\mathcal{W}$.

For the proof we recall that each coroot H_α belongs to a sub Lie algebra $\mathfrak{g}^{(\alpha)} = ((H_\alpha, X_\alpha, X_{-\alpha}))$ of type A_1 (§2.5). Applying A_1-representation theory (§1.12) to the restriction of φ to $\mathfrak{g}^{(\alpha)}$ we conclude that the operator $\varphi(H_i)$ is diagonizable. All the various $\varphi(H_i)$ commute. It is a standard result of linear algebra that then there is a simultaneous diagonalization of all the $\varphi(H_\alpha)$. This proves (a), since the H_α span $\mathfrak{h}$. Point (b) is also immediate, since by our A_1-results all eigenvalues of $\varphi(H_\alpha)$, i.e. the $\lambda(H_\alpha)$ for all the weights λ, are integers.

The proofs for (c) and (d) are a bit more elaborate: Let v be a weight vector, with weight λ, and let α be a root of $\mathfrak{g}$. Because of $H_{-\alpha} = -H_\alpha$ we may assume $\lambda(H_\alpha) > 0$ (the case $= 0$ being trivial).

Applying Lemma A to $X_{-\alpha}$ and iterating, we find that $(X_{-\alpha})^r v$, if not 0, is weight vector to the weight $\lambda - r\alpha$. But it follows from the nature of the reps D_s of A_1 that, with $r = \lambda(H_\alpha)$ (= the eigenvalue of H_α for v), the vectors v, $X_{-\alpha}v, (X_{-\alpha})^2 v, ..., (X_{-\alpha})^r v$ are non-zero, in fact independent. This proves (c) (note $\epsilon = 1$ at present). The argument shows at the same time that $m_\lambda \le m_{S_\alpha\lambda}$ (namely, the map $(X_{-\alpha})^r$ is injective on V_λ). Since S_α is an involution, we have equality here, and then $m_\lambda = m_{S\lambda}$ follows for all S in $\mathcal{W}$.

The last argument also shows $m_\lambda \le m_{\lambda-\alpha}$, provided $\lambda(H_\alpha) > 0$. Thus the sequence $m_\lambda, m_{\lambda-\alpha}, m_{\lambda-2\alpha}, \ldots, m_{S_\alpha\lambda}$ increases (weakly) up to its middle, and decreases (weakly) in the second half.

The multiplicities m_λ may well be greater than 1. This happens, e.g., for the adjoint representation, where the weight 0 appears with multiplicity l (the rank of $\mathfrak{g}$). (The other weights are the roots, with multiplicities 1.) $\surd$

Remark: (c) implies that the integers k for which $\lambda + k\alpha$ is a weight of φ fill out some interval $[-r, s]$ in $\mathbb{Z}$, with $r, s \geq 0$; these weights form the *α-string of λ* (for φ). Thus the set of weights of φ is "convex in direction α".

A weight λ of φ is *extreme* (or *highest*) if $\lambda + \alpha$ is not weight of φ for any positive root α; note that this involves the given weak order in $\mathfrak{h}_0^\top$. Extreme weights exist: we can simply take a maximal weight of φ in the given order, or we can take any weight of maximal norm (wr to the Killing form) and transform it into the closed fundamental Weyl chamber by some element of $\mathcal{W}$ (then $\langle \lambda, \alpha \rangle \geq 0$ for all α in Δ^+ and so $|\lambda + \alpha| > |\lambda|$, so that $\lambda + \alpha$ is not a weight). Similarly a weight vector of φ is called *extreme* if it is sent to 0 by the operators X_α for all positive roots α. We note an important consequence of Lemma A: *A weight vector v whose weight λ is extreme is itself extreme.*

The main construction for representation theory, generalizing directly that for A_1, follows now: Let v be an extreme weight vector of φ, with weight λ (like the vector v_0 for A_1-theory, an eigenvector of H and sent to 0 by X_+, see §1. 12). We associate to v the subspace V_v of V defined as the smallest subspace of V that contains v and is invariant under all the root elements X_{-i} corresponding to the negatives of the fundamental roots α_i. Clearly V_v is spanned by all vectors of the form $X_{-i_1}X_{-i_2}...X_{-i_k}v$ with $k = 0, 1, 2, ...$ and $1 \leq i_j \leq l$. (Thus we have v itself, all $X_{-i}v$, all $X_{-j}X_{-i}v$, etc., analogous to the vectors $v_0, X_-v_0, (X_-)^2v_0$, ... of A_1-theory.) By Lemma A each such vector, if not 0, is weight vector of φ with weight $\lambda - \alpha_{i_1} - \alpha_{i_2} - ... - \alpha_{i_k}$; it follows that all but a finite number of these vectors are 0.

PROPOSITION C. *V_v is a $\mathfrak{g}$-invariant subspace of V.*

For the proof we note that $\mathfrak{g}$ is generated by the (fundamental) root elements X_i and X_{-i} (see §2.11). Therefore it is enough to show that V_v is invariant under the X_i and X_{-i}. Invariance under the X_{-i} is part of the definition of V_v. Invariance under the X_i we prove by induction: Writing I for a sequence $\{i_1, i_2, ..., i_k\}$ as above, we abbreviate $X_{-i_1}X_{-i_2}...X_{-i_k}v$ to $X_I v$ (so $X_{\{i\}}v = X_{-i}v$); call k the *length* of I. We shall prove inductively that all $X_I v$ with I of length at most any given t are sent into V_v by the X_i.

This is clear for $t = 0$, since v is an extreme vector: all $X_i v$ are 0. For the induction, take any $k \leq t + 1$; put $I' = \{i_2, ..., i_k\}$ (with I as above). Then from the "commutation relation" $[X_i X_{-j}] = X_i X_{-j} - X_{-j} X_i$ we have $X_i X_I v = X_i X_{-i_1} X_{I'} v = X_{-i_1} X_i X_{I'} v + [X_i X_{-i_1}] X_{I'} v$. By induction the vector $X_i X_{I'} v$ is in V_v, and so is then its X_{-i_1}-image, taking care of the first term on the right. As for the second term, $[X_i X_{-i_1}]$ is 0 if i_1 is different from i (since $\alpha_i - \alpha_{i_1}$ is not a root), and is H_i if $i_1 = i$; in the latter case $X_{I'}v$ is eigenvector of H_i. $\surd$

COROLLARY D. *If the representation φ is irreducible, then there exists exactly one extreme weight, say λ; it is dominant (belongs to the semigroup $\mathcal{I}^d$), maximal in the given order, of maximal norm, and of multiplicity 1; all other weights are of the form $\lambda - \sum n_i \alpha_i$ with non-negative integers n_i.*

Proof: We take any extreme weight λ and the corresponding weight vector v (as noted, these exist). The corresponding space V_v is then invariant and non- zero and by irreducibility equals the whole space V. The claim about the uniqueness and multiplicity of λ and the form of the other weights follow at once from the explicit description of the vectors $X_I v$ spanning V_v. The other properties of λ follow by uniqueness from the fact that, as noted above, extreme weights with these properties exist. $\surd$

We interpolate a convexity property of the set of weights of φ_λ.

PROPOSITION E. *The set of weights of φ_λ is contained in the convex closure of the orbit $\mathcal{W} \cdot \lambda$ of λ under the Weyl group.*

Proof: Let μ be a weight; we may assume μ in the closed dual fundamental chamber $C^{\top -}$. From $\mu = \lambda - \sum n_i \alpha_i$ we conclude $\lambda(H) \geq \mu(H)$ for all H in C^- (i.e. with all $\alpha(H) \geq 0$). Now we apply Prop. I of §2. 11. $\surd$

We return to the situation of Cor.D. The principal fact of representation theory, which we prove below, is that conversely the extreme weight determines the representation; if two irreps of $\mathfrak{g}$ have the same extreme weight, then they are equivalent (uniqueness). Moreover, every λ in $\mathcal{I}^d$ appears as extreme weight of some irrep (existence). Clearly this gives a very good hold on the irreps. And for general, reducible reps there is Weyl's theorem that any rep is direct sum of irreps. We state these results formally:

THEOREM F. *Assigning to each irrep its extreme weight sets up a bijection between the set $\mathfrak{g}^\wedge$ of equivalence classes of irreps of $\mathfrak{g}$ and the set $\mathcal{I}^d$ of dominant integral forms in $\mathfrak{h}_0^\top$.*

THEOREM G. *Every representation of $\mathfrak{g}$ is completely reducible.*

Comments: The bijection in Theorem F seems to depend on the choice of order in $\mathfrak{h}_0^\top$ or of the fundamental Weyl chamber. One can free it from this choice be replacing the dominant weight λ in question by its orbit under the Weyl group $\mathcal{W}$, which has exactly one element in every closed Weyl chamber by Prop. H in §2.11. The bijection is then between the set $\mathfrak{g}^\wedge$ and the set of $\mathcal{W}$-orbits in the lattice $\mathcal{I}$ of integral forms.

The splitting of a rep φ into irreps, given by Theorem G, is not quite unique (if there are multiplicities, i.e., if several of the irreps are equivalent).

What is unique, is the splitting into *isotypic summands*, where such a summand is a maximal invariant subspace all of whose irreducible subspaces are $\mathfrak{g}$-isomorphic to each other. This follows easily from Schur's Lemma; an isotypic subspace is simply, in a given splitting into irreps, the sum of the spaces of all those irreps that are equivalent to a given one.

3.3 Uniqueness and existence

We start with the uniqueness part of Theorem F, the easy part. Let φ and φ' be two irreps of $\mathfrak{g}$, on the vector spaces V and V', with the same extreme weight λ. We must show φ and φ' equivalent.

The clue is the consideration of the direct sum representation $\varphi \oplus \varphi'$ on $V \oplus V'$. Let v and v' be extreme weight vectors to λ for φ and φ'; then (v, v') clearly is an extreme weight vector to λ for $\varphi \oplus \varphi'$, with associated invariant subspace $W = (V \oplus V')_{(v,v')}$ (see Prop. C in §2). The (equivariant) projection p of $V \oplus V'$ onto V sends (v, v') to v, and therefore (by irreducibility) maps W onto V. On the other hand the kernel of p on W is the intersection of W with the natural summand V of $V \oplus V'$, and thus a $\mathfrak{g}$-invariant subspace of V. It cannot contain the vector v, since (v, v') is the only vector in W with weight λ (all the vectors generated from (v, v') have lower weights). Thus by irreducibility of φ this kernel is 0, and so p is an equivariant isomorphism of V with W. Similarly W is isomorphic to V', and so V and V' are isomorphic, i.e., φ and φ' are equivalent. $\checkmark$

We come to the hard part, existence of irreps. The proof we give is an ad hoc version of the standard proof (which involves the *Poincaré-Birkhoff-Witt theorem*, the *Borel sub Lie algebra* of $\mathfrak{g}$ (spanned by $\mathfrak{h}$ and the X_α for all positive α), and the *Verma module* (similar to our V^λ below)).

Let λ be a dominant integral form on $\mathfrak{h}_0^\top$. We must construct an irrep φ with λ as extreme weight. We shall construct, successively: First an infinite dimensional vector space U^λ on which the elements X_α for α in Δ and the H in $\mathfrak{h}$ act (but brackets are not preserved; this is not quite a representation of $\mathfrak{g}$); U^λ will be a direct sum of finite dimensional eigenspaces of $\mathfrak{h}$ with weights in $\mathcal{I}$ of the form "λ minus a sum of positive roots", and with λ an extreme weight of multiplicity 1. Second, a quotient V^λ of U^λ, still infinite-dimensional, but otherwise with the same properties, on which the original action becomes a representation of $\mathfrak{g}$. Finally a quotient W^λ of V^λ, irreducible under $\mathfrak{g}$, with λ as extreme weight, and of finite dimension. We take our clue from the form of the space V_v in Prop. B, §3.2.

Let $\{\gamma_1, ..., \gamma_m\}$ be a list of all positive roots of $\mathfrak{g}$ (this is not a fundamental system). To each finite sequence $I = \{i_1, ..., i_k\}$ of k integers i_r satisfying $1 \le i_r \le m$, with k (the *length* of I)$= 0, 1, 2, \ldots$, we assign an abstract element v_I. Thus we have $v_\emptyset$ (also written just v), $v_1, v_2, ..., v_m, v_{11}, v_{12}, v_{21}$,

... . We let U^λ be the vector space over $\mathbb{C}$ with all these v_I as basis. For any such I and any i with $1 \le i \le m$ we put $iI = \{i, i_1, ..., i_k\}$.

We shall now define operators $\underline{H}$ for H in $\mathfrak{h}$ and $\underline{X}_\alpha$ for α in Δ, operating on U^λ; here $\underline{H}$ depends linearly on H. For an arbitrary $X = H + \sum c_\alpha X_\alpha$ we then put $\underline{X} = \underline{H} + \sum c_\alpha \underline{X}_\alpha$; thus $\underline{X}$ is linear in X.

We define $\underline{H}$ as follows: Any v_I, with I as above, is eigenvector of $\underline{H}$ with eigenvalue $\lambda(H) - \gamma_{i_1}(H) - \gamma_{i_2}(H) - ... - \gamma_{i_k}(H)$. Clearly U^λ is then direct sum of weight spaces U^λ_μ with weights of the form $\mu = \lambda - \gamma_{i_1} - \gamma_{i_2} - ... - \gamma_{i_k}$. Each such weight space is of finite dimension, because of the positivity of the γ_i. Clearly also the various $\underline{H}$ commute with each other; we have a representation of $\mathfrak{h}$.

Next, for any γ_i we define $\underline{X}_{-\gamma_i}$ in the obvious way: $\underline{X}_{-\gamma_i} v_I = v_{iI}$. Finally we define $\underline{X}_{\gamma_i} v_I$ by induction on the length of I: To begin with we put $\underline{X}_{\gamma_i} v = 0$. We denote the operator assigned to $[X_\alpha X_\beta]$ by $\underline{X}_{\alpha\beta}$ for any α, β in Δ; this equals $\underline{H}_\alpha$ if $\beta = -\alpha$, or $N_{\alpha\beta} \underline{X}_{\alpha+\beta}$ if $\alpha + \beta$ is a root, and the operator 0 otherwise. (Recall that we put $N_{\lambda\mu} = 0$, if one of $\lambda, \mu, \lambda + \mu$ is not a root, and similarly $\underline{X}_\sigma = 0$ for any σ in $\mathfrak{h}_0^\top - \Delta$.) For any I of length > 0 we write I in the form $i_1 I'$ and put $\underline{X}_{\gamma_i} v_I (= \underline{X}_{\gamma_i} \underline{X}_{-\gamma_{i_1}} v'_I) = \underline{X}_{-\gamma_{i_1}} \underline{X}_{\gamma_i} v_{I'} + \underline{X}_{\gamma_i, -\gamma_{i_1}} v_{I'}$. (Note that the operations on $v_{I'}$ are already defined inductively.) Thus we are forcing $\underline{X}_{\gamma_i} \underline{X}_{-\gamma_j} - \underline{X}_{-\gamma_j} \underline{X}_{\gamma_i} = \underline{X}_{\gamma_i, -\gamma_j}$.

With α, β in Δ we write $\underline{Z}_{\alpha\beta}$ for $\underline{X}_\alpha \underline{X}_\beta - \underline{X}_\beta \underline{X}_\alpha - \underline{X}_{\alpha\beta}$, and define $\underline{Z}_{\lambda\mu}$ to mean the operator 0 for λ, μ in $\mathfrak{h}_0^\top$, but at least one of λ, μ not a root; note that the relations $\underline{Z}_{\alpha\beta} = 0$ hold for $\alpha > 0, \beta < 0$ and for $\alpha < 0, \beta > 0$, but possibly not for the remaining cases. $\underline{H}\underline{X}_\alpha - \underline{X}_\alpha \underline{H}$ is the operator to $[HX_\alpha]$, i.e. it equals $\alpha(H)\underline{X}_\alpha$, for all α, from the easily verified fact that $\underline{X}_\alpha$ sends a vector of weight ρ to one of weight $\rho + \alpha$. To force $\underline{Z}_{\alpha\beta} = 0$ for all pairs of roots α, β and thus to get a representation of $\mathfrak{g}$, we form the smallest subspace, say U', of U^λ that contains all $\underline{Z}_{\alpha\beta} v_I$ and is invariant under all operators $\underline{X}_\alpha$ and $\underline{H}$. It is fairly clear that U' is spanned by all vectors of the form $\underline{X}_{\delta_1} \underline{X}_{\delta_2} \cdots \underline{X}_{\delta_k} \underline{Z}_{\alpha\beta} v_I$, with the δ_i in Δ.

On the quotient space $V^\lambda = U^\lambda / U'$ we have then induced operators X'_α and H', and generally X', which form a representation of $\mathfrak{g}$, since the relations $X'_\alpha X'_\beta - X'_\beta X'_\alpha = [X_\alpha X_\beta]'$ now hold for all α and β. Furthermore, V^λ is spanned by the images of the v_I (which we still call v_I; they may not be independent any more), and so v generates V^λ under the action of $\mathfrak{g}$. The v_I are eigenvectors of the H', with the same eigenvalues as before. V^λ is still direct sum of (finite dimensional) weight spaces of $\mathfrak{h}$. (This uses a standard argument of linear algebra, essentially the same as the one showing that eigenvectors of an operator to different eigenvalues are linearly independent.) In particular λ is an extreme weight, of multiplicity 1, with v as eigenvector, provided v is not 0 in V^λ (this proviso is equivalent to $V^\lambda \ne 0$ or $U' \ne U^\lambda$).

Thus, in order to get something non-trivial, we must show that the vector v (in U^λ) does not belong to U'. Since v is the only basis vector of weight λ, this amounts to the following.

LEMMA A. *Let α and β or $-\alpha$ and $-\beta$ be in Δ^+. Then for arbitrary δ_i and ϵ_j in Δ with $\sum\delta_i+\alpha+\beta+\sum\epsilon_j=0$ the vector $\underline{X}_{\delta_1}\cdots\underline{X}_{\delta_+}\cdot \underline{Z}_{\alpha\beta}\cdot\underline{X}_{\epsilon_1}\cdots\underline{X}_{\epsilon_s}v$ is 0.*

(By the relation on the δ_i and ϵ_j the vector in the lemma is of weight λ.) We start the proof with two auxiliary relations.

(*) If α,β,γ are in Δ^+, then
$$\underline{Z}_{\alpha\beta}\cdot\underline{X}_{-\gamma}=\underline{X}_{-\gamma}\cdot\underline{Z}_{\alpha\beta}+N_{\alpha,-\gamma}\underline{Z}_{\alpha-\gamma,\beta}+N_{\beta,-\gamma}\underline{Z}_{\alpha,\beta-\gamma}.$$

(**) If α,β,γ are in Δ^+, then
$$\underline{X}_\gamma\cdot\underline{Z}_{-\alpha,-\beta}=\underline{Z}_{-\alpha,-\beta}\cdot\underline{X}_\gamma+N_{\gamma,-\alpha}\cdot\underline{Z}_{\gamma-\alpha,-\beta}+N_{\gamma,-\beta}\underline{Z}_{-\alpha,\gamma-\beta}.$$

Proof of (*): Using $\underline{Z}_{\beta,-\gamma}=0$ etc., we get

$$\begin{aligned}
&\underline{Z}_{\alpha\beta}\cdot\underline{X}_{-\gamma}\\
&\quad= \underline{X}_\alpha\cdot\underline{X}_\beta\underline{X}_{-\gamma}-\underline{X}_\beta\cdot\underline{X}_\alpha\underline{X}_{-\gamma}-N_{\alpha\beta}\underline{X}_{\alpha+\beta}\underline{X}_{-\gamma}\\
&\quad= \underline{X}_\alpha\cdot(\underline{X}_{-\gamma}\underline{X}_\beta+N_{\beta,-\gamma}\underline{X}_{\beta-\gamma})-\underline{X}_\beta\,(\underline{X}_{-\gamma}\underline{X}_\alpha+\\
&\qquad N_{\alpha,-\gamma}\underline{X}_{\alpha-\gamma})-N_{\alpha\beta}\,(\underline{X}_{-\gamma}\underline{X}_{\alpha+\beta}+N_{\alpha+\beta,-\gamma}\underline{X}_{\alpha+\beta-\gamma})\\
&\quad= (\underline{X}_{-\gamma}\underline{X}_\alpha+N_{\alpha,-\gamma}\underline{X}_{\alpha-\gamma})\,\underline{X}_\beta+N_{\beta,-\gamma}\underline{X}_\alpha\underline{X}_{\beta-\gamma}\\
&\qquad-(\underline{X}_{-\gamma}\underline{X}_\beta+N_{\beta,-\gamma}\underline{X}_{\beta-\gamma})\,\underline{X}_\alpha-N_{\alpha,-\gamma}\underline{X}_\beta\underline{X}_{\alpha-\gamma}\\
&\qquad-N_{\alpha\beta}\underline{X}_{-\gamma}\underline{X}_{\alpha+\beta}-N_{\alpha\beta}N_{\alpha+\beta,-\gamma}\underline{X}_{\alpha+\beta-\gamma}\,.
\end{aligned}$$

Here the term $N_{\beta,-\gamma}\underline{X}_{\beta-\gamma}$ should be replaced by $\underline{H}_\beta$, if $\beta=\gamma$; similarly for $\gamma=\alpha$ or $\gamma=\alpha+\beta$.

Equation (*) follows upon applying the relation $N_{\alpha\beta}N_{\alpha+\beta,-\gamma}= N_{\beta,-\gamma}N_{\alpha,\beta-\gamma}+N_{\alpha,-\gamma}N_{\beta,\alpha-\gamma}$, which follows from the Jacobi identity for X_α, X_β, and $X_{-\gamma}$ or the vanishing of some N's; again this has to be modified if $\gamma=\alpha$ (replace the last term by $\beta(H_\alpha)$) or β or $\alpha+\beta$. Similarly for (**). √

We can now prove Lemma A: We apply (*) and (**), and also the relations $\underline{X}_\theta\underline{X}_{-\eta}=\underline{X}_{-\eta}\underline{X}_\theta+\underline{X}_{\theta,-\eta}$ (i.e., $\underline{Z}_{\theta,-\eta}=0$) for $\theta,\eta>0$, to the vector in the lemma, in order to shift all factors $\underline{X}_\delta$ and $\underline{X}_\epsilon$ with δ_i or $\epsilon_j<0$ all the way to the left, in the case $\alpha,\beta>0$, or to shift the $\underline{X}_\delta$ and $\underline{X}_\epsilon$ with δ_i or $\epsilon_j>0$ all the way to the right, in the case $\alpha,\beta<0$.

These shifts introduce additional, similar (with other $\underline{Z}$'s), but shorter terms (i.e., smaller s or t), which are 0 by induction assumption. After the shifts have been completed, the term is 0: In case $\alpha,\beta>0$ it must begin with at least one $\underline{X}_{-\gamma}$; but v is not in the image space of any such operator, by definition. In case $\alpha,\beta<0$ the first operator applied to v must be an $\underline{X}_\gamma$; but those operators annul v. The induction starts with terms as in Lemma A that do not allow any of our shifts. But then the vector in question is 0, by the argument just given. √

We now have $V^\lambda = U^\lambda/U'$, with a representation of $\mathfrak{g}$ on it. As noted earlier, it is direct sum of finite dimensional weight spaces. The same argument shows that this holds also for any $\mathfrak{g}$-invariant (or $\mathfrak{h}$-invariant) subspace. Therefore among all $\mathfrak{g}$-invariant proper subspaces (i.e., different from V^λ itself, or, equivalently, not containing v) there is a unique maximal one. Dividing V^λ by it, we get a quotient space W^λ with an irreducible representation of $\mathfrak{g}$ on it, still generated by v under the action of $\mathfrak{g}$, with λ as extreme weight, and direct sum of finite dimensional weight spaces. We continue to write the spanning vectors as v_I. We plan to show that W^λ has finite dimension—which will establish the existence theorem.

We recall that for $i = 1, \ldots, l$ we have the fundamental roots α_i, the coroots H_i, the root elements X_i and X_{-i}, and the sub Lie algebras $\mathfrak{g}^{(i)} = ((H_i, X_i, X_{-i}))$ of $\mathfrak{g}$ (which are of type A_1, with $[H_iX_i] = 2X_i, [H_iX_{-i}] = -2X_{-i}, [X_iX_{-i}] = H_i$). The X_i's and X_{-i}'s generate $\mathfrak{g}$. We prove two lemmas.

LEMMA B. *For each i from 1 to l the space W^λ is sum of finite-dimensional $\mathfrak{g}^{(i)}$-invariant subspaces.*

We fix i and show first that there exists a non-trivial finite-dimensional $\mathfrak{g}^{(i)}$-invariant subspace: We consider the sequence $w_0 = v, w_1 = X_{-i}w_0$, $w_2 = X_{-i}w_1, \ldots$. The computations of A_1-theory (§1.11) yield the relations $X_iw_t = \mu_t w_{t-1}$ with $\mu_t = t(r - t + 1)$, where $r = \lambda(H_i)$ is a non-negative integer. We see that X_iw_{r+1} is 0. For $j \neq i$ we get $X_jw_{r+1} = X_j(X_{-i})^{r+1}v = (X_{-i})^{r+1}X_jv$, since X_j and X_{-i} commute (α_i and α_j being fundamental, $\alpha_j - \alpha_i$ cannot be a root), and so $X_jw_{r+1} = 0$. Thus w_{r+1} is an extreme vector, and the computation for Proposition C in §2 shows that the space generated from w_{r+1} by the X_{-i} is $\mathfrak{g}^{(i)}$-invariant. This space is clearly not the whole space W^λ (all weights are less than λ), and so by irreducibility of W^λ it is 0. In particular w_{r+1} is 0. It follows that the space $((w_0, w_1, \ldots, w_r))$ is $\mathfrak{g}^{(i)}$-invariant; so non-zero finite-dimensional $\mathfrak{g}^{(i)}$-invariant subspaces exist.

Next we note: If U is a finite-dimensional $\mathfrak{g}^{(i)}$-invariant subspace of W^λ, so is the space $\mathfrak{g}U$ generated by U under $\mathfrak{g}$, i.e., the space spanned by all Xu with X in $\mathfrak{g}$ and u in U, because of $X_{\pm i}Xu = XX_{\pm i}u + [X_{\pm i}X]u$. Therefore the span of all finite-dimensional $\mathfrak{g}^{(i)}$-invariant subspaces is $\mathfrak{g}$-invariant. It is not 0, as shown above, and thus by irreducibility it is equal to W^λ. $\surd$

LEMMA C. *The set of weights that occur in W^λ is invariant under the Weyl group.*

Proof: Let θ be a weight of W^λ, with weight vector w. Take any i between 1 and l; we have to show that $S_i\theta$, i.e. $\theta - \theta(H_i)\alpha_i$, is also a weight. By lemma B and by A_1-theory the vector w lies in a finite direct sum of $\mathfrak{g}^{(i)}$-invariant subspaces in which certain of the standard irreducible reps D_s

appear. Suppose $\theta(H_i) > 0$ (a similar argument works if $\theta(H_i)$ is negative; the case $\theta(H_i) = 0$ is trivial). We write r for the positive integer $\theta(H_i)$, and note that w is eigenvector of H with eigenvalue r. We know from A_1-theory that $w' = (X_{-i})^r w$ is then eigenvector of H_i (with eigenvalue $-r$), and in particular it is not 0. But Lemma B of §3.2 tells us that w' is weight vector with weight $\theta - r\alpha_i$, and so $\theta - \theta(H_i)\alpha_i$ is a weight of W^λ. $\surd$

We come now to the main fact, which finally establishes the existence of a finite-dimensional representation of $\mathfrak{g}$ with extreme weight λ.

PROPOSITION D. *The dimension of W^λ is finite.*

Clearly it is enough to show that W^λ has only a finite number of weights; by Lemma C it is enough to show that W^λ has only a finite number of dominant weights, i.e. in the closed fundamental Weyl chamber $C^{\top -}$. That this holds, comes from the simple geometric fact that the half space $\{\sigma \in \mathfrak{h}_0^\top : \sigma \leq \lambda\}$ intersects $C^{\top -}$ in a bounded set. In detail: All the weights μ in question are of the form $\Sigma n_i \lambda_i$ (where the λ_i are the fundamental weights and the n_i are non-negative integers); they also satisfy $\mu \leq \lambda$ (since they are of the form λ minus a sum of positive roots). But there is only a finite number of integral forms with these two properties: Let H_0 be the element of $\mathfrak{h}_0$ that defines the order. The λ_i are positive, by Proposition A of §3.1, so we have $\lambda_i(H_0) > 0$. The condition $\lambda \geq \mu$ translates into $\lambda(H_0) \geq \Sigma n_i \lambda_i(H_0)$. Clearly this leaves only a finite number of possibilities for the n_i. $\surd$

With this Theorem F of §3.2 is proved.

(Note: To the weight $\lambda = 0$ corresponds of course the trivial representation.)

3.4 Complete reduction

We prove Theorem G of §3.2. Let φ be a representation of $\mathfrak{g}$ on V (irreducible or not). We recall the notion of trace form t_φ of φ (§1.5): $t_\varphi(X,Y) = \mathrm{tr}\,(\varphi(X) \cdot \varphi(Y))$. (Also recall our use of Xv for $\varphi(X)(v)$. We will even write X for $\varphi(X)$ and depend on the context to determine whether X is meant in $\mathfrak{g}$ or in $\mathfrak{gl}(V)$.)

LEMMA A. *If φ is faithful, then the trace form t_φ is non-degenerate.*

For the proof we consider the set $\mathfrak{j} = \{X \in \mathfrak{g} : t_\varphi(X,Y) = 0$ for all Y in $\mathfrak{g}\}$, the *radical* of t_φ. By infinitesimal invariance of t_φ (loc.cit.) this is an ideal in $\mathfrak{g}$. By assumption we may consider $\mathfrak{g}$ as a sub Lie algebra of $\mathfrak{gl}(V)$. Proposition B, §1.9 says then that $\mathfrak{j}$ is solvable; by semisimplicity of $\mathfrak{g}$ it must be 0. $\surd$

Next comes an important construction, the *Casimir operator* Γ_φ of φ: Let $\mathfrak{a}$ be the (unique) ideal of $\mathfrak{g}$ complementary to $\ker\varphi$; by restriction φ defines a faithful representation of $\mathfrak{a}$. Let $X_1, \ldots, X_n$ be any basis for $\mathfrak{a}$, let $Y_1, \ldots, Y_n$ be the dual basis wr to the trace form on $\mathfrak{a}$ (so that $t_\varphi(X_i, Y_j) = \delta_{ij}$), and put $\Gamma_\varphi = \Sigma\varphi(X_i) \circ \varphi(Y_i)$. It is easily verified that this is independent of the choice of the basis $\{X_i\}$. The basic properties of Γ_φ appear in the next proposition and corollary.

PROPOSITION B.

(a) Γ_φ *commutes with all operators* $\varphi(X)$*;*

(b) $\operatorname{tr}(\Gamma_\varphi) = \dim\mathfrak{a} = \dim\mathfrak{g} - \dim\ker\varphi$.

Proof: Take any X in $\mathfrak{g}$. We expand $[XX_i]$ (which lies in $\mathfrak{a}$) as $\Sigma x_{ij}X_j$ and $[XY_i]$ as $\Sigma y_{ij}Y_j$. We have $x_{ij} = \operatorname{tr}[XX_i]Y_j$, and the latter equals $-\operatorname{tr}X_i[XY_j] = -y_{ji}$, by invariance of t_φ (§1.5). Then we compute $[X\Gamma_\varphi] = \Sigma[XX_i]Y_i + \Sigma X_i[XY_i] = \Sigma x_{ij}X_jY_i + \Sigma y_{ij}X_iY_j$
$= 0$, proving (a). And (b) is immediate from $\operatorname{tr}X_iY_i = 1$. √

COROLLARY C. *If* φ *is irreducible (and* $V \neq 0$*), then* Γ_φ *is the scalar operator* $(\dim\mathfrak{g} - \dim\ker\varphi)/\dim V \cdot id$*; it is thus non-singular, if* φ *is non-trivial* .

Proof: By part (a) of Proposition B and Schur's lemma the operator Γ_φ is scalar; the value of the scalar follows from part (b). √

The key to complete reducibility is the next result, known as JHC Whitehead's first lemma. ("The cohomology space $H^1(\mathfrak{g}, V)$ is 0.")

PROPOSITION D. *Let* $\mathfrak{g}$ *act on* V *(as above). Let* $f : \mathfrak{g} \to V$ *be a linear function satisfying the relation* $f([XY]) = Xf(Y) - Yf(X)$ *for all* X, Y *in* $\mathfrak{g}$*. Then there exists a vector* v *in* $\mathfrak{g}$ *with* $f(X) = Xv$ *for all* X *in* $\mathfrak{g}$.

(Note that for given v the function $X \to Xv$ satisfies the relation that appears in Proposition D, which is thus a necessary condition.)

Proof: First suppose that V has an invariant subspace U, with quotient space W and quotient map $\pi : V \to W$. We show: If Proposition D holds for U and W, it also holds for V. Let w in W satisfy $\pi \cdot f(X) = Xw$, let w' be a representative for w in V, and define the function f' by $X \to f(X) - Xw'$. We have $\pi \cdot f'(X) = 0$ for all X, i.e., f' maps $\mathfrak{g}$ into U. Also, f' has the property of Proposition D. Therefore there is a u in U with $f'(X) = Xu$ for all X. But this means $f(X) = X(w' + u)$ for all X, and Proposition D holds for V.

Thus we have to prove Proposition D only for irreducible V. This is trivial for the trivial rep ($\dim V = 1$, all $X = 0$). Suppose then φ is irreducible and non-trivial, so that by Corollary C the Casimir operator Γ_φ is invertible. As in the case of Proposition B, let $\{X_i\}$ and $\{Y_i\}$ be dual bases of $\mathfrak{a}$ wr to t_φ. We define v in V by the equation $\Gamma_\varphi(v) = \Sigma X_i f(Y_i)$. Then we have $\Gamma_\varphi(Xv - f(X)) = \Sigma X X_i f(Y_i) - \Sigma X_i Y_i f(X)) = \Sigma[XX_i]f(Y_i) + \Sigma X_i(Xf(Y_i) - Y_i f(X)) + = \Sigma[XX_i]f(Y_i) + \Sigma X_i f([XY_i]) = \Sigma x_{ij} X_j f(Y_i) + \Sigma y_{ij} X_i f(Y_j) = 0$ for all X, and so $f(X) = Xv$ for all X. $\surd$

We come now to complete reducibility and prove Theorem G of §3.2.

So let $\mathfrak{g}$ act on V, via φ, let U be an invariant subspace, and let W be the quotient space, with quotient map $\pi : V \to W$. We have to find a complementary invariant subspace, or, equivalently, we have to find a $\mathfrak{g}$-equivariant map of W into V, whose composition with π is id_W.

We write L and M for the vector spaces of all linear maps of W into U and V. (We can think of L as a subspace of M.) There is an action of $\mathfrak{g}$ on these two spaces, defined for X in $\mathfrak{g}$ by $p \to [Xp] = X \cdot p - p \cdot X$ (this makes sense for any linear map p between two $\mathfrak{g}$-spaces). The equivariant maps are the invariants of this action, i.e., those with $[Xp] = 0$ for all X in $\mathfrak{g}$. Let $\mathfrak{h}$ be any element of M with $\pi \cdot h = id_W$ (this exists since π is surjective). We plan to make $\mathfrak{h}$ equivariant by subtracting a suitable element of L.

Consider the map $X \to [Xh]$, a map f of $\mathfrak{g}$ into M that satisfies the relation in Proposition D (see the remark after Proposition D). The composition $\pi \cdot [Xh] = \pi \cdot X \cdot h - \pi \cdot h \cdot X$ is 0, by $\pi \cdot X = X \cdot \pi$ and $\pi \cdot h = id_W$, for any X. This means that $[Xh]$ actually lies in L, so f can be considered as a map of $\mathfrak{g}$ into L. We apply JHC Whitehead's lemma (Proposition D) to it as such: There exists a k in L with $f(X) = [Xk]$. Thus we have $[X, (h-k)] = 0$ for all X, i.e., $h - k$ is an equivariant map of W into V; and the relation $\pi \cdot k = 0$ (from $\pi(U) = 0$) shows $\pi \cdot (h-k) = \pi \cdot h = id_W$. So $h - k$ does what we want. $\surd$

We have now finished the proof of the main result, Theorems F and G of §3.2, existence and uniqueness of the irrep to prescribed dominant weight λ.

One might of course consider reps of real semisimple Lie algebras. Complex representations are the same as those of the complexification; so there is nothing new. We shall not go into the considerations needed for classifying real, real-irreducible reps. Complete reduction goes through for real reps almost exactly as in the complex case. The only difference is that the Casimir operator for an irrep is not necessarily scalar (as in Corollary C); it is however still non-0 (since its trace is not 0) and thus invertible (by Schur's lemma), and that is enough for the argument.

For completeness's sake we sketch the proof of a related result.

THEOREM E. *Let $\mathfrak{g}$ be the direct sum of two (semisimple) Lie algebras $\mathfrak{g}_1$ and $\mathfrak{g}_2$. Then any irrep φ of $\mathfrak{g}$ is (equivalent to) the tensor product* *of two irreps* φ_1 *and* φ_2 *of* $\mathfrak{g}_1$ *and* $\mathfrak{g}_2$.

This reduces the representations of a semisimple Lie algebra to those of its simple summands. In terms of our main results, it will be clear that the extreme weight of φ in Theorem E is the sum of the extreme weights of φ_1 and φ_2.

Proof: Let φ' be the restriction of φ to the summand φ_1. By complete reducibility V splits into the direct sum of some φ'-invariant-and-irreducible subspaces $V_1, V_2, \ldots$. All the V_i are isomorphic as $\mathfrak{g}_1$-spaces: Since $\mathfrak{g}_1$ and $\mathfrak{g}_2$ commute, the map $V_1 \to V_i$ obtained by operating with any Y in $\mathfrak{g}_2$ and then projecting into V_i is $\mathfrak{g}_1$-equivariant and therefore, by Schur's lemma, an isomorphism or 0; the sum of the V_i that are $\mathfrak{g}_1$-isomorphic to V_1 (an isotypic component of V) is $\mathfrak{g}$-invariant and so equal to V. Thus we can write V as $V_1 \oplus V_1 \oplus \cdots \oplus V_1$, as $\mathfrak{g}_1$-space, or (writing φ_1 for the action of $\mathfrak{g}_1$ on V_1) also as $V_1 \otimes W$ with X in $\mathfrak{g}_1$ acting as $\varphi_1(X) \otimes id$, with a suitable space W.

Take any Y in $\mathfrak{g}_2$. As before, the map of the i-th summand V_1 obtained by first operating with $\varphi(Y)$ and then projecting to the j-th summand is $\mathfrak{g}_1$-equivariant and therefore scalar. Interpreted in the form $V_1 \otimes W$ of V this means that there is a representation φ_2 of $\mathfrak{g}_2$ on W with $\varphi(Y) = id \otimes \varphi_2(Y)$. Clearly φ_2 has to be irreducible, and $\varphi(X, Y)$ is $\varphi_1(X) \otimes id + id \otimes \varphi_2(Y)$. $\checkmark$

The converse is also true (over $\mathbb{C}$): If φ_1, φ_2 are irreps of $\mathfrak{g}_1$, $\mathfrak{g}_2$, then $\varphi \otimes \varphi_2$ is an irrep of $\mathfrak{g}_1 \otimes \mathfrak{g}_2$.

As an application we look at the irreps of the Lorentz Lie algebra $\mathfrak{l}_{3,1}$ (Example 11, §1.1). We recall from §1.4 that it is isomorphic to $\mathfrak{sl}(2, \mathbb{C})_{\mathbb{R}} = (A_1)_{\mathbb{R}}$. Its complexification is $A_1 \oplus A_1^-$, and the irreps of the latter are the tensorproducts $D_s \otimes D_t^-$ with $s, t \in \{0, 1/2, 1, 3/2, \ldots\}$. Restricting to $\mathfrak{l}_{3,1}$ (as it sits in $A_1 \oplus A_1^-$) and spelling out what the D_s are, we find the (complex) irreps $D_{s,t}$ of the Lorentz Lie algebra (i.e., of $(A_1)_{\mathbb{R}}$) as tensorproducts of the space of homogeneous polynomials in ξ and η of degree s and the space of homogeneous polynomials in the complex conjugate variables ξ^- and η^- of degree t (each matrix in A_1 acting via its complex-conjugate).

3.5 Cartan semigroup; representation ring

Let $\mathfrak{g}$ be semisimple as before; we continue with the previous notations etc.

The set $\mathcal{D}$ of all (equivalence classes of) representations (not necessarily irreducible) of $\mathfrak{g}$ is a semiring, with direct sum and tensor product as sum and product. To get an actual ring out of this, one introduces the *representation ring*, also *character ring* or *Grothendieck ring of virtual representations* $R\mathfrak{g}$:

Write $[\varphi]$ for the equivalence class of the rep φ. The additive group of $R\mathfrak{g}$ is simply the universal (Abelian) group attached to the additive group of $\mathcal{D}$ (cf. $\mathbb{Z}$ and $\mathbb{N}$): We consider pairs $([\varphi], [\psi])$ of representation classes (which eventually will become differences $[\varphi] - [\psi]$), with componentwise addition, and call two pairs $([\varphi], [\psi]), ([\varphi'], [\psi'])$ eqivalent if $\varphi \oplus \psi'$ is equivalent to $\varphi' \oplus \psi$. Then $R\mathfrak{g}$, additively, is the set of equivalence classes of these pairs, with the induced addition. The tensor product of reps induces a product in $R\mathfrak{g}$, under which it becomes a (commutative) ring. (The trivial rep becomes the unit.) One writes $[\varphi]$ for $([\varphi], [0])$ (in $R\mathfrak{g}$); $([0], [\psi])$ then becomes $-[\psi], ([\varphi], [\psi])$ becomes $[\varphi] - [\psi]$, and $[\varphi \oplus \psi]$ equals $[\varphi] + [\psi]$.

(It is because of the appearance of minus signs that one speaks of virtual representations. An integral-linear combination of reps represents 0 if the direct sum of the terms with positive coefficients is equivalent to that of the terms with negative coefficients.) (Note: We used tacitly that complete reduction implies cancelation; i.e., $\varphi \oplus \psi_1 \approx \varphi \oplus \psi_2$ implies $\psi_1 \approx \psi_2$. Otherwise one would have to define equivalence of pairs by: there exists χ with $\varphi \oplus \psi' \oplus \chi \approx \varphi' \oplus \psi \oplus \chi$.)

For an alternate description, write (temporarily) F for the free Abelian group generated by the set $\mathcal{D}$ and let N be the subgroup of F generated by all elements of the form $[\varphi \oplus \psi] - [\varphi] - [\psi]$; then the additive group of $R\mathfrak{g}$ is by definition the quotient group F/N, and multiplication is induced by the tensor product. Equivalence of the two definitions comes, e.g., from the universal property: Every additive map of $\mathcal{D}$ into any Abelian group A extends uniquely to a homomorphism of $R\mathfrak{g}$ into A. One also sees easily that additively $R\mathfrak{g}$ is a free Abelian group with the set $\mathfrak{g}^\wedge$ of (classes of) irreps as basis. ($\mathfrak{g}^\wedge$ generates $R\mathfrak{g}$ by the complete reducibility theorem. The map of F that sends each rep into the sum of its irreducible constituents vanishes on N and thus factors through $R\mathfrak{g}$, and shows that there are no linear relations between the elements of $\mathfrak{g}^\wedge$ in $R\mathfrak{g}$.)

Consider two irreps φ and φ' of $\mathfrak{g}$, on vector spaces V and V', with extreme weights λ and λ'. The tensor product rep $\varphi \otimes \varphi'$ of $\mathfrak{g}$, on $V \otimes V'$, is not necessarily irreducible (in fact, it is almost always reducible). (Note that as a rep of $\mathfrak{g} \oplus \mathfrak{g}$ it would be irreducible, but that in effect we are restricting this rep to the "diagonal" sub Lie algebra of $\mathfrak{g} \oplus \mathfrak{g}$, the set of pairs (X, X).) By complete reducibility it splits then into a certain number of irreps. In §3.8 we shall give a "formula" for this splitting (cf. the Clebsch-Gordan series of §1.12); but for the moment we have a less ambitious goal.

Let v, v' be weight vectors of φ, φ', with weights ρ, ρ'; it is clear from the definition of $\varphi \otimes \varphi'$ that $v \otimes v'$ is weight vector of $\varphi \otimes \varphi'$, with weight $\rho + \rho'$, and that one gets all weight vectors and weights of $\varphi \otimes \varphi'$ this way. In particular, since λ and λ' have multiplicity 1, $\lambda + \lambda'$ is the unique maximal weight of $\varphi \otimes \varphi'$ (thus extreme) and it has multiplicity 1. This means that in the decomposition of $\varphi \otimes \varphi'$ the irrep with extreme weight $\lambda + \lambda'$ occurs exactly once, and that all other irreps that occur have smaller extreme weight.

The irrep with extreme weight $\lambda + \lambda'$ is called the *Cartan product* of φ and φ'. The set $\mathfrak{g}^\wedge$ of equivalence classes of irreps of $\mathfrak{g}$, endowed with this product, is called the *Cartan semigroup* (of irreps of $\mathfrak{g}$). It is now clear from the main result (Theorem E in §3.3) that assigning to each irrep its extreme weight sets up an isomorphism between the Cartan semigroup $\mathfrak{g}^\wedge$ and the (additive) semigroup $\mathfrak{I}^d$ of dominant weights. We recall that $\mathfrak{I}^d$ is generated (freely) by the fundamental weights $\lambda_1, \ldots, \lambda_l$; the corresponding irreps are called the *fundamental* reps and denoted by $\varphi_1, \ldots, \varphi_l$.

The structure of the Cartan semigroup has a strong consequence for the structure of the representation ring $R\mathfrak{g}$.

THEOREM A. *The ring $R\mathfrak{g}$ is isomorphic (under the natural map) to the polynomial ring $P_{\mathbf{Z}}[\varphi_1, \ldots, \varphi_l]$ in the fundamental reps φ_i.*

In other words, the φ_i generate $R\mathfrak{g}$, and there are no linear relations between the various monomials in the φ_i. For the obvious natural homomorphism Ψ of the polynomial ring into $R\mathfrak{g}$ we first prove surjectivity by induction wr to the order in $\mathfrak{h}_o^\top$. Let $\lambda = \Sigma n_i \lambda_i$ be a dominant weight, and assume that all the elements of $\mathfrak{g}^\wedge$ with smaller extreme weight are in the image of Ψ.

We form $\varphi_1^{n_1} \otimes \ldots \otimes \varphi_l^{n_l}$ (the exponents are meant in the sense of tensor product), the Ψ-image of the monomial $\varphi_1^{n_1} \ldots \varphi_l^{n_l}$ in the polynomial ring. By the discussion above this is the sum of the irrep φ_λ belonging to λ and other terms that belong to lower extreme weights. Since all the other terms belong to the image of Ψ already, so does φ_λ. $\surd$

Next injectivity of Ψ. For a given non-zero polynomial we pick out a monomial whose associated weight $\lambda = \Sigma n_i \lambda_i$ is maximal. The argument just used shows that the Ψ-image of the polynomial in $R\mathfrak{g}$ involves the irrep φ_λ with a non-zero coefficient (the other monomials can't interfere), and so is not 0. We shall return to this topic in §3.7. $\surd$

3.6 The simple Lie algebras

We now turn to the simple Lie algebras. Using the notation developed in §2.13 we shall list for each type the fundamental coroots H_i and the translation lattice $\mathcal{T}$, the fundamental weights λ_i, the lowest form δ, and the fundamental reps φ_i. For completeness we also describe the center lattice $\mathcal{Z}$, and the *connectivity group* $\mathcal{Z}/\mathcal{T}$.

If φ is a representation of $\mathfrak{g}$, on a vector space V, we write $\varphi \wedge \varphi$ or $\bigwedge^2 \varphi$ for the induced representation on the exterior product $\bigwedge^2 V$, and more generally $\bigwedge^r \varphi$ for the induced representation on the r-th exterior power $\bigwedge^r V = V \wedge V \wedge \ldots \wedge V$. (In more detail: $\varphi \wedge \varphi(X)$ sends $v \wedge w$ to $Xv \wedge$

$w + v \wedge Xw$.) If $\rho_1, \rho_2, \ldots$ are the weights of φ (possibly with repetitions), with weight vectors $v_1, v_2, \ldots$, then the $\rho_i + \rho_j$ with $i < j$ are the weights of $\varphi \wedge \varphi$, with the $v_i \wedge v_j$ as weight vectors; more generally the weights on $\bigwedge^r V$ are the sums $\rho_{i_1} + \rho_{i_2} + \ldots + \rho_{i_r}$ with $i_1 < i_2 < \ldots < i_r$ and with the corresponding products $v_{i_1} \wedge v_{i_2} \wedge \ldots \wedge v_{i_r}$ as weight vectors. As usual we write e_i for the i–th coordinate vector in $\mathbb{R}^n$ or $\mathbb{C}^n$ (thus $e_1 = (1, 0, \ldots, 0)$ etc.), and ω_i for the i–th coordinate function. — An analogous description holds for the induced rep on the symmetric products $S^r V$.

1) A_l, $\mathfrak{sl}(l+1, \mathbb{C})$.

(Recall the restriction $\omega_1 + \omega_2 + \ldots + \omega_{l+1} = 0$ for $\mathfrak{h}$; elements of $\mathfrak{h}^\top$ are linear combinations of $\omega_1, \ldots, \omega_{l+1}$ modulo the term $\Sigma_1^{l+1} \omega_i$.)

$H_1 = e_1 - e_2, H_2 = e_2 - e_3, \ldots, H_l = e_l - e_{l+1}$.
$\lambda_1 = \omega_1, \lambda_2 = \omega_1 + \omega_2, \ldots, \lambda_l = \omega_1 + \omega_2 + \ldots + \omega_l$.
$\delta = l \cdot \omega_1 + (l-1) \cdot \omega_2 + \ldots + 1 \cdot \omega_l$.
$\varphi_1 = \mathfrak{sl}(l+1, \mathbb{C}) =$ the representation of $\mathfrak{sl}(l+1, \mathbb{C})$ "by itself", $= \Lambda_1$ in short, $\varphi_2 = \bigwedge^2 \mathfrak{sl}(l+1, \mathbb{C}) = \Lambda_2, \ldots, \varphi_l = \bigwedge^l \mathfrak{sl}(l+1, \mathbb{C}) = \Lambda_l$.

$\mathcal{T}$: The $H = (a_1, a_2, ,,, a_{l+1})$ with all coordinates a_i integral (and of course $\Sigma a_i = 0$).
$\mathcal{Z}$: The H such that for some integer k all a_i are congruent to $k/l+1$ mod 1 (and $\Sigma a_i = 0$).
$\mathcal{Z}/\mathcal{T} = \mathbb{Z}/l+1$ (the cyclic group of order $l+1$).
$\mathcal{I}^d$ (the dominant forms): the forms $\lambda = \Sigma_1^l f_i \omega_i$ with integral f_i satisfying $f_1 \geq f_2 \geq \ldots \geq f_l \geq 0$.

To justify these statements we recall that the Killing form on $\mathfrak{h}$ is the restriction to the subspace $\omega_1 + \ldots + \omega_{l+1} = 0$ of $\mathbb{C}^{l+1}$ of the usual Euclidean form $\Sigma_1^{l+1} \omega_i^2$, up to a factor. Therefore the root vector h_{12} corresponding to the root $\alpha_{12} = \omega_1 - \omega_2$ is certainly proportional to $e_1 - e_2$; and since the latter vector has the correct value 2 on α_{12}, it is the coroot H_{12}.

The λ_i exhibited are clearly the dual basis to the H_i; we have $\lambda_i(H_j) = \delta_{ij}$. The conditions that define $\mathcal{I}^d$ simply say that the values $\lambda(H_i)$ are non-negative integers. Note that λ in reality is of the form $\Sigma_1^{l+1} f_i \omega_i$ and is defined only mod $\Sigma_1^{l+1} \omega_i$, and that in effect we have normalized λ by putting $f_{l+1} = 0$.

The weights of Λ^1 are the ω_i, $i = 1, 2, \ldots, l+1$, since $\mathfrak{h}$ consists of the diagonal matrices (of trace 0). The weights of Λ^r are the $\omega_{i_1} + \omega_{i_2} + \ldots + \omega_{i_r}$ with $1 \leq i_1 < i_2 < \cdots < i_r \leq l+1$. This is the orbit of λ_r under the Weyl group (all permutations of the coordinates). Since the irrep φ_r to λ_r as extreme weight must have all these as weights, it follows that Λ_r is φ_r. That $\mathcal{T}$ is as described is fairly clear from the form of the H_i. For $\mathcal{Z}$ note that all roots $\omega_i - \omega_j$ are integral on an H in $\mathcal{Z}$; i.e., all a_i are congruent to each other mod 1 (and $\Sigma a_i = 0$). For $\mathcal{Z}/\mathcal{T}$: The vector

$v_1 = (1/l+1, \ldots, 1/l+1, -l/l+1)$ and its multiples $2v_1, \ldots, (l+1)v_1$ form a complete system of representatives of $\mathcal{Z}$ mod $\mathcal{T}$.

[For the general linear Lie algebra $\mathfrak{gl}(l+1, \mathbb{C})$ - which is not semisimple (it is $\mathfrak{sl}(l+1, \mathbb{C}) \oplus \mathbb{C}$, where the second term is the one-dimensional, Abelian, center) - the situation is as follows: We have the irreps $\Lambda_i = \bigwedge^i \mathfrak{gl}(l+1, \mathbb{C})$ for $i = 1, 2, \ldots, l+1$. The last one, Λ_{l+1}, (which didn't appear for $\mathfrak{sl}$) is one-dimensional and assigns to each matrix its trace (on the group level this is the map matrix $\rightarrow$ determinant). The tensor power $(\Lambda_{l+1})^n$ makes sense for all integral n, even negative ones (matrix $\rightarrow$ n·trace). [Formally one could consider matrix $\rightarrow$ c·trace for any constant c; but in order to get single-valued reps for the corresponding general linear group one must take c integral.] The notion of weight etc. makes sense roughly as for $\mathfrak{sl}$ (but the reps in question should have their restriction to the center completely reducible, i.e., $\varphi(id)$ should be a diagonalizable matrix), with $\mathfrak{h}$ now being the set of all diagonal matrices. In this sense one has an irrep for each weight of the form $\Sigma_1^{l+1} n_i \lambda_i$ with $n_i \geq 0$ for $i = 1, 2, \ldots, l$, but with n_{l+1} running through all of $\mathbb{Z}$ (here λ_{l+1} means of course $\omega_1 + \omega_2 + \ldots + \omega_{l+1}$). One can change n_{l+1} by tensoring with a tensor power of Λ_{l+1}. The representation ring is $P\mathbb{Z}[\Lambda_1, \Lambda_2, \ldots, \Lambda_{l+1}, (\Lambda_{l+1})^{-1}]$, i.e. $P_{\mathbb{Z}}[\Lambda_1, \ldots \Lambda_{l+1}, x]$ mod the ideal generated by $x \cdot \Lambda_{l+1} - 1$.]

We shall give less detail in the remaining cases.

2) $B_l, = \mathfrak{o}(2l+1, \mathbb{C})$.

$H_1 = e_1 - e_2, \ldots, H_{l-1} = e_{l-1} - e_l, H_l = 2e_l$.
$\lambda_1 = \omega_1, \lambda_2 = \omega_1 + \omega_2, \ldots, \lambda_{l-1} = \omega_1 + \omega_2 + \cdots + \omega_{l-1}, \lambda_l = 1/2(\omega_1 + \omega_2 + \cdots + \omega_l)$.
$\delta = (l - 1/2)\omega_1 + (l - 3/2)\omega_2 + \cdots + 1/2\omega_l$
$\varphi_1 = \mathfrak{o}(2l+1, \mathbb{C}) = \Lambda_1, \varphi_2 = \bigwedge^2 \varphi_1 = \Lambda_2, \ldots, \varphi_{l-1} = \Lambda_{l-1}$; finally φ_l corresponding to the "unusual" weight λ_l, is a quite "non-obvious" representation, called the *spin representation* and denoted by Δ_l or just Δ, of dimension 2^l as we shall see in the next section. (The proper algebraic construction for the spin rep is through *Clifford algebras.*)

$\mathcal{T}$: The H with integral coordinates a_i and even Σa_i.
$\mathcal{Z}$: The H with all a_i integral.
$\mathcal{Z}/\mathcal{T} = \mathbb{Z}/2$; e_1 is a representative of the non-trivial element.
$\mathcal{I}^d$ consists of the forms $\lambda = \Sigma_1^l f_i \omega_i$ with $f_1 \geq f_2 \geq \cdots \geq f_l \geq 0$, all f_i integral, or all f_i half-integral (i.e., congruent to $1/2$ mod 1). (These conditions express again the integrality of the $\lambda(H_i)$.)

3) $C_l, = \mathfrak{sp}(l, \mathbb{C})$.

$H_1 = e_1 - e_2, \ldots, H_{l-1} = e_{l-1} - e_l, H_l = e_l$.
$\lambda_i = \omega_1 + \omega_2 + \cdots + \omega_i$ for $i = 1, 2, \ldots, l$.
$\delta = l \cdot \omega_1 + (l-1) \cdot \omega_2 + \cdots + \omega_l$.

φ_1 is again Λ_1, $= \mathfrak{sp}(l, \mathbb{C})$ itself, on $\mathbb{C}^{2l}$. For the other φ_i : The basic 2-form Ω of $\mathbb{C}^{2l}$ maps $\bigwedge^i \mathbb{C}^{2l}$ onto $\bigwedge^{i-2} \mathbb{C}^{2l}$ ("inner product" or contraction, dual to the map $\bigwedge^{i-2}(\mathbb{C}^{2l})^\top$ to $\bigwedge^i(\mathbb{C}^{2l})^\top$ by exterior product with Ω). Since Ω is invariant under $\mathfrak{sp}(l, \mathbb{C})$, the map is equivariant, and its kernel is an invariant subspace. The restriction Λ_i of $\bigwedge^i \varphi_1$ to this kernel is φ_i, for $i = 2, \ldots, l$. (With coordinates: Represent Ω by the skew matrix $[a_{rs}]$; a skew tensor $t^{u_1 u_2 \ldots u_m}$ goes to $a_{rs} t^{u_1 u_2 \ldots u_{m-2} rs}$.)

$\mathcal{T}$: The H with all a_i integral
$\mathcal{Z}$: The H with all a_i integral or all half-integral ($\equiv 1/2 \bmod 1$)
$\mathcal{Z}/\mathcal{T} = \mathbb{Z}/2$.
$\mathcal{I}^d$ consists of the $\Sigma f_i \omega_i$ with f_i integral, $f_1 \geq f_2 \geq \cdots \geq f_l \geq 0$.

4) $D_l, = \mathfrak{o}(2l, \mathbb{C})$.

$H_1 = e_1 - e_2, \ldots, H_{l-1} = e_{l-1} - e_l, H_l = e_{l-1} + e_l$.
$\lambda_i = \omega_1 + \omega_2 + \ldots + \omega_i$ for $1 \leq i \leq l-2$,
$\lambda_{l-1} = 1/2(\omega_1 + \omega_2 + \ldots + \omega_{l-1} - \omega_l), \lambda_l = 1/2(\omega_1 + \omega_2 + \cdots + \omega_{l-1} + \omega_l)$.
$\delta = (l-1)\omega_1 + (l-2)\omega_2 + \cdots + \omega_{l-1}$.
$\varphi_1 = \mathfrak{o}(2l, \mathbb{C}) = \Lambda_1, \varphi_2 = \bigwedge^2 \varphi_1 = \Lambda_2, \ldots, \varphi_{l-2} = \bigwedge^{l-2} \varphi_1 = \Lambda_{l-2}$.
In addition there are two non-obvious irreps, called the *negative and positive half-spin representations*, $\varphi_{l-1} = \Delta_l^-$ and $\varphi_l = \Delta_l^+$; both are of dimension 2^{l-1}, as we shall see in the next section. (Again the proper context is Clifford algebras.)

$\mathcal{T}$: The H with integral a_i and even Σa_i.
$\mathcal{Z}$: The H with all a_i integral or all a_i half-integral.
$\mathcal{Z}/\mathcal{T} = \mathbb{Z}/4$ for odd l, $\mathbb{Z}/2 \oplus \mathbb{Z}/2$ for even l.
The point $P = (1/2, 1/2, \ldots, 1/2)$ is a representative for a generator for odd l; P and $Q = (1/2, 1/2, \ldots, 1/2, -1/2)$ are representatives for the generators of the two $\mathbb{Z}/2\mathbb{Z}$'s for even l.
$\mathcal{I}^d$ consists of the $\Sigma f_i \omega_i$ with the f_i all integral or all half-integral and $f_1 \geq f_2 \geq \ldots \geq f_{l-1} \geq |f_l|$. (Note the absolutevalue.)

5) G_2.

$H_1 = (1, -1, 0), H_2 = (-1, 2, -1)$.
$\lambda_1 = \omega_1 - \omega_3, \lambda_2 = \omega_1 + \omega_2$. (Recall $\omega_1 + \omega_2 + \omega_3 = 0$.)
$\delta = 3\omega_1 + 2\omega_2$.
φ_1 has dimension 14; it is the adjoint representation. φ_2 has dimension 7; it identifies G_2 with the Lie algebra of derivations of the (eight-dimensional algebra of) Cayley numbers, or rather with its complexification (see [12]).

$\mathcal{T}$: The $H = (a_1, a_2, a_3)$ with integral a_i and $a_1 + a_2 + a_3 = 0$.
$\mathcal{Z} = \mathcal{T}$
$\mathcal{Z}/\mathcal{T} = 0$.

$\mathcal{I}^d$: The $\Sigma f_i\omega_i$ with the differences between the f_i integral and with $f_1 \geq f_2, 2f_2 \geq f_1 + f_3$. (Or, making use of $\omega_1 + \omega_2 + \omega_3 = 0$, the $f_1\omega_1 + f_2\omega_2$ with f_1 and f_2 integral and $2f_2 \geq f_1 \geq f_2$.)

6) F_4.

$H_1 = e_1 - e_2 - e_3 - e_4$, $H_2 = 2e_4, H_3 = e_3 - e_4, H_4 = e_2 - e_3$.
$\lambda_1 = \omega_1$, $\lambda_2 = 1/2(3\omega_1 + \omega_2 + \omega_3 + \omega_4)$, $\lambda_3 = 2\omega_1 + \omega_2 + \omega_3, \lambda_4 = \omega_1 + \omega_2$.
$\delta = 1/2(11\omega_1 + 5\omega_2 + 3\omega_3 + \omega_4)$.

Center lattice $\mathcal{Z}$: The H with integral a_i and even Σa_i.
$\mathcal{T} = \mathcal{Z}$.
$\mathcal{Z}/\mathcal{T}$ trivial.
$\mathcal{I}^d$: The $\Sigma f_i\omega_i$ with the f_i all integral or all half-integral, and with $f_2 \geq f_3 \geq f_4 \geq 0$ and $f_1 \geq f_2 + f_3 + f_4$.

7) E_6.

($\mathfrak{h}$ as described in §2.14.)

$H_1 = e_1 - e_2, \ldots, H_5 = e_5 - e_6, H_6 = 1/3(-e_1 - e_2 - e_3 + 2e_4 + 2e_5 + 2e_6)$.
$\lambda_1 = 1/3(4\omega_1 + \omega_2 + \cdots + \omega_6)$, $\lambda_2 = 1/3(5\omega_1 + 5\omega_2 + 2\omega_3 + \cdots + 2\omega_6)$, $\lambda_3 = 2(\omega_1 + \omega_2 + \omega_3) + \omega_4 + \omega_5 + \omega_6, \lambda_4 = 4/3(\omega_1 + \cdots + \omega_4) + 1/3(\omega_5 + \omega_6)$, $\lambda_5 = 2/3(\omega_1 + \cdots + \omega_5) - 1/3\omega_6, \lambda_6 = \omega_1 + \cdots + \omega_6$.
$\delta = 8\omega_1 + 7\omega_2 + 6\omega_3 + 5\omega_4 + 4\omega_5 + 3\omega_6$.

Center lattice $\mathcal{Z}$: The H with $a_i, i > 0$, all integral or all $\equiv 1/3 \mod 1$ or all $\equiv 2/3 \mod 1$.
Coroot lattice $\mathcal{T}$: The sublattice of $\mathcal{Z}$ with $4a_1 + a_2 + \cdots + a_6 \equiv 0 \mod 3$.
$\mathcal{Z}/\mathcal{T} = \mathbb{Z}/3$. A representative for a generator is e_1.
$\mathcal{I}^d$: The $\Sigma_1^6 f_i\omega_i$ with $3f_i$ integral, all differences $f_i - f_j$ integral, $f_1 + f_2 + f_3 - 2(f_4 + f_5 + f_6)$ integral and divisible by 3, $f_1 \geq f_2 \geq \cdots \geq f_6$ and $f_1 + f_2 + f_3 \leq 2(f_4 + f_5 + f_6)$.

8) E_7.

($\mathfrak{h}$ as described in §2.14.)

$H_1 = e_1 - e_2, \ldots, H_6 = e_6 - e_7, H_7 = 1/3(-e_1 - \cdots - e_4 + 2e_5 + 2e_6 + 2e_7)$.
$\lambda_1 = 1/2(3\omega_1 + \omega_2 + \cdots + \omega_7)$, $\lambda_2 = 2(\omega_1 + \omega_2) + \omega_3 + \cdots + \omega_7$, $\lambda_3 = 5/2(\omega_1 + \omega_2 + \omega_3) + 3/2(\omega_4 + \cdots + \omega_7)$, $\lambda_4 = 3(\omega_1 + \cdots + \omega_4) + 2(\omega_5 + \omega_6 + \omega_7)$, $\lambda_5 = 2(\omega_1 + \cdots + \omega_5) + \omega_6 + \omega_7$, $\lambda_6 = \omega_1 + \cdots + \omega_6$, $\lambda_7 = 3/2(\omega_1 + \cdots + \omega_7)$.
$\delta = 1/2(27\omega_1 + 25\omega_2 + 23\omega_3 + 21\omega_4 + 19\omega_5 + 17\omega_6 + 15\omega_7)$.

Center lattice $\mathcal{Z}$: The H with a_i, all integral or all $\equiv 1/3 \mod 1$ or all $\equiv 2/3 \mod 1$.
Coroot lattice $\mathcal{T}$: The sublattice of $\mathcal{Z}$ with $3a_1 + a_2 + \cdots + a_7 \equiv 0 \mod 2$.
$\mathcal{Z}/\mathcal{T} = \mathbb{Z}/2$. A representative for the generator is e_1.

$\mathcal{I}^d$: The $\Sigma_1^7 f_i\omega_i$ with the f_i all integral or all half-integral, $2\Sigma_1^7 f_i$ divisible by 3, $f_1 \geq f_2 \geq \ldots \geq f_7$ and $f_1 + f_2 + f_3 + f_4 \leq 2(f_5 + f_6 + f_7)$.

9) E_8.

(Subspace $\sum_1^9 \omega_i = 0$ of $\mathbb{C}^9$ as in §2.14.)

$H_1 = e_1 - e_2, \ldots, H_7 = e_7 - e_8, H_8 = 1/3(-e_1 - \cdots - e_5 + 2e_6 + 2e_7 + 2e_8 - e_9)$
$\lambda_i = \omega_1 + \cdots + \omega_i - i\omega_9$ for $1 \leq i \leq 5$. $\lambda_6 = 2/3(\omega_1 + \cdots + \omega_6) - 1/3(\omega_7 + \omega_8) - 10/3\omega_9$, $\lambda_7 = 1/3(\omega_1 + \cdots + \omega_7) - 2/3\omega_8 - 5/3\omega_9$, $\lambda_8 = 1/3(\omega_1 + \cdots + \omega_8) - 8/3\omega_9$.
$\delta = 1/3(19\omega_1 + 16\omega_2 + 13\omega_3 + 10\omega_4 + 7\omega_5 + 4\omega_6 + \omega_7 - 2\omega_8 - 68\omega_9)$.

Center lattice $\mathcal{Z}$: The H with the a_i all integral or all $\equiv 1/3 \bmod 1$ or all $\equiv 2/3 \bmod 1$, and $a_i = 0$.
Coroot lattice $\mathcal{T} = \mathcal{Z}$.
$\mathcal{Z}/\mathcal{T}$ trivial.

$\mathcal{I}^d$: The $\Sigma_1^9 f_i\omega_i$ with the f_i all integral or all $\equiv 1/3 \bmod 1$ or all $\equiv 2/3 \bmod 1$, $\Sigma f_i = 0$, $f_1 \geq f_2 \geq \cdots \geq f_8$ and $f_6 + f_7 + f_8 \geq 0$.

(In the second picture for E_8, with $\mathfrak{h} = \mathbb{C}^8$, we have
$H_1 = 1/2\sum e_i, H_2 = -e_1 - e_2, H_3 = e_2 - e_3, H_4 = e_1 - e_2, H_5 = e_3 - e_4, H_6 = e_4 - e_5, H_7 = e_5 - e_6, H_8 = e_6 - e_7$. $\lambda_1 = 2\omega_8, \lambda_2 = 1/2(-\omega_1 - \omega_2 - \cdots - \omega_7 + 7\omega_8), \lambda_3 = -\omega_3 - \cdots - \omega_7 + 5\omega_8, \lambda_4 = 1/2(\omega_1 - \omega_2 - \omega_3 - \cdots - \omega_7 + 5\omega_8), \lambda_5 = -\omega_4 - \cdots - \omega_7 + 4\omega_8, \lambda_6 = -\omega_5 - \omega_6 - \omega_7 + 3\omega_8, \lambda_7 = -\omega_6 - \omega_7 + 2\omega_8, \lambda_8 = -\omega_7 + \omega_8$. $\delta = -\sum_1^7 (i-1)\omega_i + 23\omega_8$.

Center lattice $\mathcal{Z}$, = coroot lattice $\mathcal{T}$: The $\Sigma a_i e_i$ with the a_i all $\equiv 0$ or all $\equiv 1/2 \bmod 1$ and the sum $\Sigma_1^8 a_i$ an even integer.
$\mathcal{I}^d$: The $\sum f_i\omega_i$ with the f_i all integral or all $\equiv 1/2 \bmod 1$ and $\sum f_i$ even.)

In Figs.3, 4, 5 we present the Cartan-Stiefel diagrams for A_2, B_2, G_2. The figures can be interpreted as $\mathfrak{h}_0^\top$ or as $\mathfrak{h}_0$.

For $\mathfrak{h}_0^\top$ the points marked □ form the lattice $\mathcal{R}$, and the points marked ○ form the lattice $\mathcal{I}$; the vectors marked α and β form a fundamental system of roots.

For $\mathfrak{h}_0$, the points marked □ form $\mathcal{T}$, the points marked ○ form $\mathcal{Z}$; the vectors marked α and β are the coroots H_β and H_α (in that order!; a long root corresponds to a short coroot).
The fundamental Weyl chamber is shaded. The fundamental weights (in $\mathfrak{h}_0^\top$) are the two points nearest the origin on the edges of the fundamental Weyl chamber. Their sum is the element δ.
Note that for G_2 one has $\mathcal{R} = \mathcal{I}$ and $\mathcal{T} = \mathcal{Z}$.

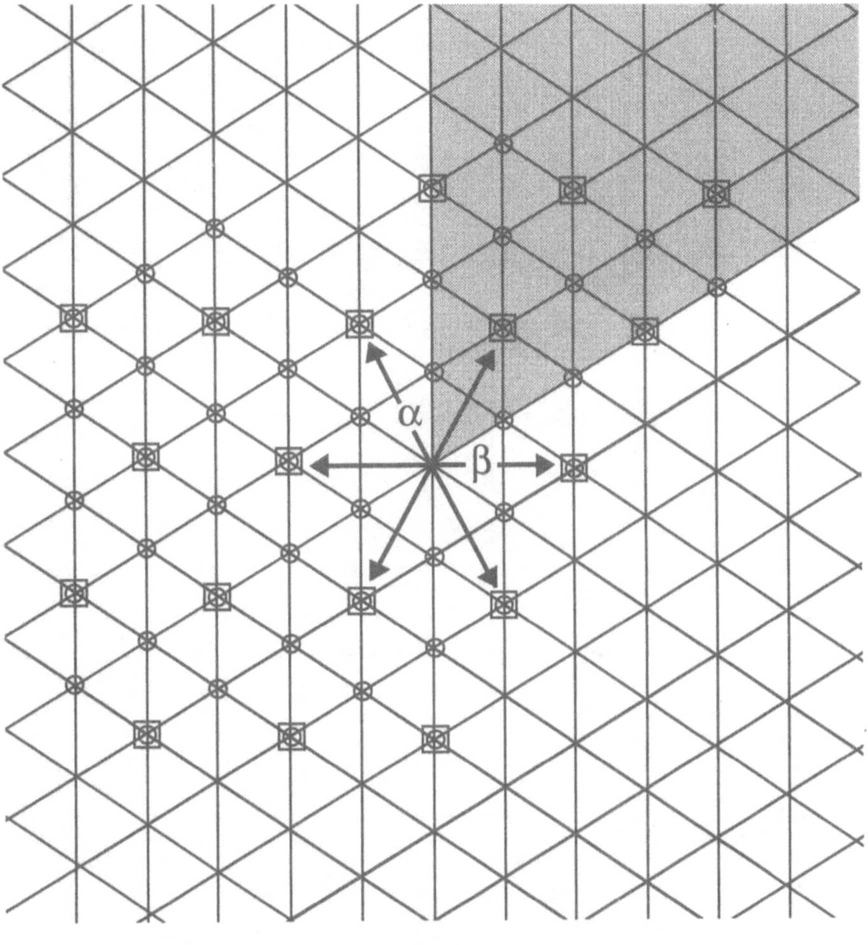

A₂

Figure 3

There exists a quite different path to the representations of the classical Lie algebras (see H.Weyl, [25]): For A_l, e.g., one starts with the "lowest" representation $\Lambda_1, \mathfrak{sl}(l+1, \mathbb{C})$ itself, forms tensor powers $(\Lambda_1)^n$ with arbitrary n, and decomposes them into irreducible subspaces by *symmetry operators*; this yields all the irreps.

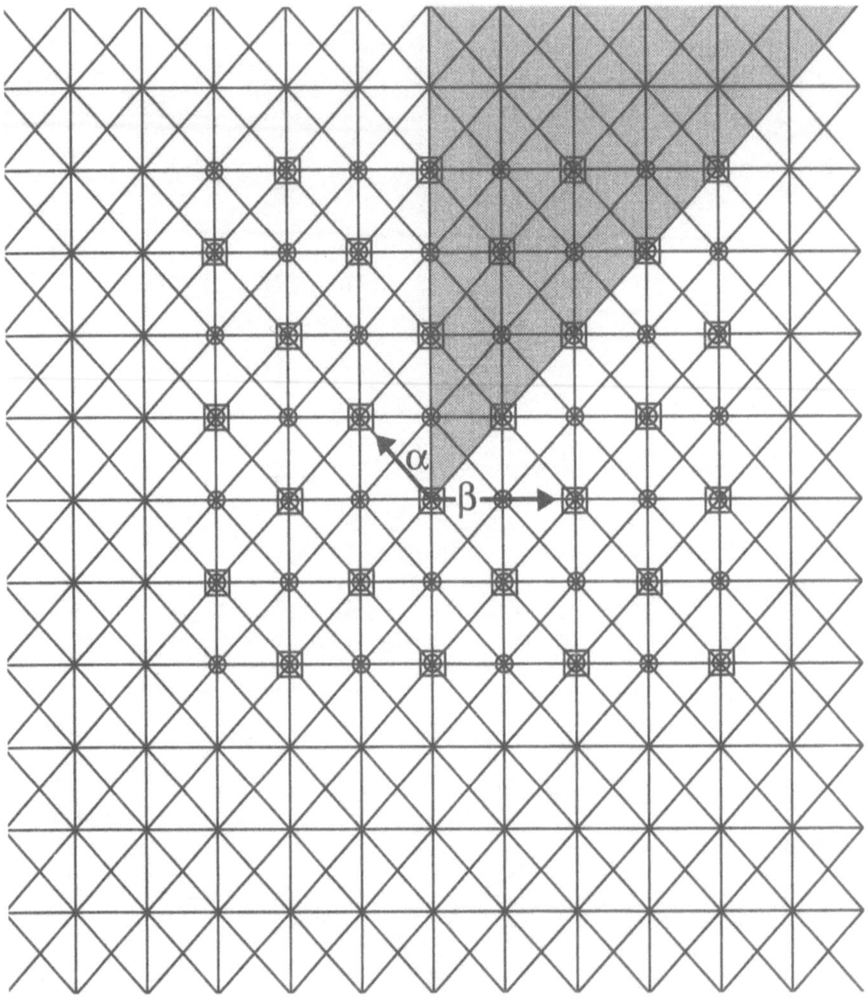

B_2

Figure 4

(NB: There are of course the two subspaces of the symmetric tensors and of the skew symmetric tensors; but there are many others.) For the other Lie algebras, B_l, C_l, D_l, one also has to put certain traces (wr to the inner or exterior product) equal to 0. (However the spin reps and the other reps of $\mathfrak{o}(n, \mathbb{C})$ with half-integral f_i do not arise this way.)

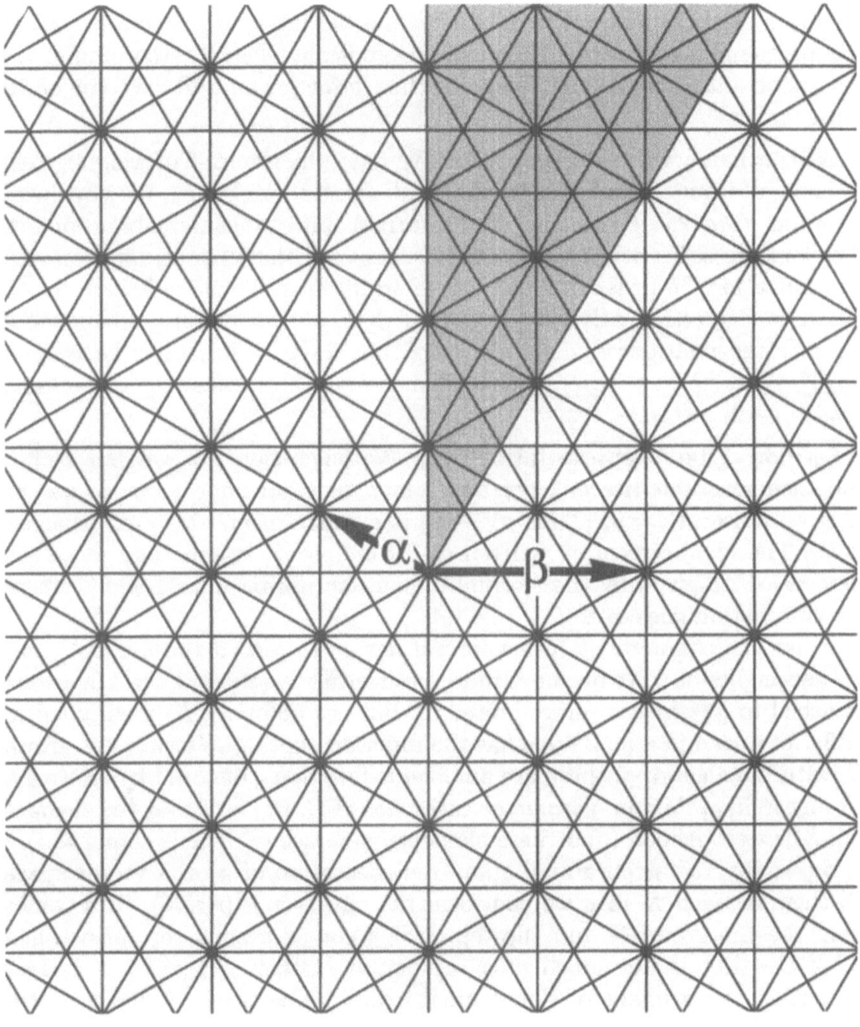

Figure 5

3.7 The Weyl character formula

We first define the concept of character of a representation φ of our Lie algebra $\mathfrak{g}$ algebraically, rather formally, and discuss it in the context of Lie groups, to make contact with the usual definition. Then we state and prove the important formula of H. Weyl for the character, and derive some of its consequences.

We continue with $\mathfrak{g}$ etc. as before. We have the group $\mathcal{I}$ of weights, free Abelian, of rank l, generated by the fundamental weights λ_i. We now form its *group ring* $\mathbb{Z}\mathcal{I}$, consisting of the formal finite linear combinations of the elements of $\mathcal{I}$ with integral coefficients, with the obvious addition and multiplication. In order not to confuse addition in $\mathcal{I}$ with addition in $\mathbb{Z}\mathcal{I}$ we write $\mathcal{I}$ multiplicatively: To each ρ in $\mathcal{I}$ we associate a new symbol e_ρ, with the relations $e_{\rho+\sigma} = e_\rho \cdot e_\sigma$. (Thus for $\rho = \Sigma n_i \lambda_i$ we have $e_\rho = (e_{\lambda_1})^{n_1} \cdot (e_{\lambda_2})^{n_2} \cdot \dots \cdot (e_{\lambda_l})^{n_l}$.) The elements of $\mathbb{Z}\mathcal{I}$ are then the finite sums $\Sigma m_\rho e_\rho$, with integers m_ρ.

Let now φ be a representation of $\mathfrak{g}$. We have then the weights ρ of φ and their multiplicities m_ρ. For any ρ in $\mathcal{I}$ that does not occur as weight of φ we write $m_\rho = 0$. [Thus $\rho \to m_\rho$ is a function $m : \mathcal{I} \to \mathbb{Z}$, attached to φ.] The *character* of φ, written as χ_φ or just χ, is now defined as the element of $\mathbb{Z}\mathcal{I}$ given by the (formal, but in fact finite) sum $\Sigma m_\rho e_\rho$, where the summation goes over $\mathcal{I}$.

So far the character is just a formal device to record the multiplicities of the weights of φ. It becomes more interesting in terms of the Lie *group*, attached to $\mathfrak{g}$ (which we have hinted at, but not defined). As mentioned in §1.3 , for any A in a $\mathfrak{gl}(V)$ one has the function $\exp(sA)$. For any Lie group G, with Lie algebra $\mathfrak{g}$, there are analogous functions, denoted by $\exp(sX)$, for any X in $\mathfrak{g}$, the *one-parameter subgroups* of G. In particular the element $\exp X$ is well defined (and these elements generate G, if G is connected).

If $\mathfrak{g}$ is any Lie algebra and φ any representation of $\mathfrak{g}$ on a vector space V, then for each X in $\mathfrak{g}$ we can form the operator $\exp \varphi(X)$. [If $\mathfrak{g}$ comes from the Lie group G and the rep φ of $\mathfrak{g}$ comes from a rep, also called φ, of G—that is not much of a restriction—, then $\exp \varphi(X)$ is in fact $\varphi(\exp X)$, the φ-image of the element $\exp X$.] The trace of this operator is a function of X, i.e., a function on $\mathfrak{g}$. [If there is a group G around as described, the value $\operatorname{tr}(\exp \varphi(X))$ equals $\operatorname{tr}(\varphi(\exp X))$, i.e., it is what is usually called the character of φ at the element $\exp X$ of G.] The standard facts continue to hold in our situation: If φ and φ' are equivalent reps, then we have $\operatorname{tr}(\exp \varphi(X)) = \operatorname{tr}(\exp \varphi'(X))$ [this is obvious]; and $\operatorname{tr}(\exp \varphi(X)) = \operatorname{tr}(\exp(\varphi(X'))$ for $X' = \exp(adY)(X)$ for any Y in $\mathfrak{g}$, analogous to the character of a group rep being constant on conjugacy classes [the relation $\varphi(X') = \exp \varphi(Y) \cdot \varphi(X) \cdot (\exp \varphi(Y))^{-1}$ holds then].

Let now $\mathfrak{g}$ be semisimple as above, with all the associated machinery. The rep φ then has its weights $\rho_1, \rho_2, \dots$, with the associated weight vectors

$v_1, v_2, \ldots$ in V. For each H in $\mathfrak{h}$ the operator $\exp(\varphi(H))$ is now diagonal, with diagonal entries $\exp(\rho_r(H))$. The character, i.e. the trace, is then of the form $\Sigma m_\rho \exp(\rho(H))$, where the sum goes over the weights of φ and the m_ρ are the multiplicities. We make one more modification by introducing a factor $2\pi i$, and define the *character* χ_φ or just χ of φ as the trace of $\exp(2\pi i\varphi(H))$, as function of H. [It makes sense to restrict oneself to $\mathfrak{h}$, since any representation is determined - up to equivalence - by its weights, which are functions on $\mathfrak{h}$. It is of course implicit that all the results below do not depend on the choice of the Cartan sub Lie algebra $\mathfrak{h}$.]

To repeat, the *character* of φ is the $\mathbb{C}$-valued function on $\mathfrak{h}$ given by $H \to \Sigma m_\rho \exp(2\pi i\rho(H))$, the sum going over $\mathcal{I}$. As a matter of fact, we will consider only the H in $\mathfrak{h}_0$ (in part the reason for this is that we can write $\exp(2\pi i\varphi(H)$ as $\exp(2\pi\varphi(iH))$, and that $i\mathfrak{h}_0$ is the Cartan sub Lie algebra of the compact form of $\mathfrak{g}$, cf. §2.10).

The main reason for the factor $2\pi i$ is that then the character, in fact every term $\exp(2\pi i\rho(H))$ in it, takes the same value at any two H's whose difference lies in the coroot lattice $\mathcal{T}$, since the ρ's are integral forms. In other words, χ is a *periodic* function on $\mathfrak{h}_0$, with the elements of $\mathcal{T}$ as periods. As usual, when dealing with functions that are periodic wr to a lattice such as $\mathcal{T}$, one considers Fourier series, with terms $c_\rho \exp(2\pi i\rho(H))$, where the ρ run over the dual lattice in the dual space - which here is of course just the lattice $\mathcal{I}$ of weights in $\mathfrak{h}_0^\top$. We see that χ is in fact a finite Fourier series. We describe this a bit differently: We form the quotient group $\mathfrak{h}_0/\mathcal{T}$ and denote it by $\mathbb{T}$. [It is isomorphic to the l-dimensional torus, i.e. $\mathbb{R}^l$ modulo the lattice of integral vectors, direct sum of l copies of $\mathbb{R}/\mathbb{Z}$. We note without proof or even explanation that $\mathbb{T}$ represents a maximal torus of the compact simply-connected Lie group associated to $\mathfrak{g}$.]

Each function $\exp \circ 2\pi i\rho$ on $\mathfrak{h}_0$, with ρ in $\mathcal{I}$, is a (continuous) homomorphism of $\mathfrak{h}_0$ into the unit circle $U = \{z : |z| = 1\}$ in $\mathbb{C}$. It has the lattice $\mathcal{T}$ in its kernel, and so induces a homomorphism of $\mathbb{T}$ into U; in the usual language for Abelian groups this is also called a character of $\mathbb{T}$ (a slightly different use of the word character). We take it as well known that we get all characters of $\mathbb{T}$ that way.

We write e_ρ for $\exp \circ 2\pi i\rho$, as function on $\mathfrak{h}_0$ (which makes sense for all ρ in $\mathfrak{h}_0^\top$) or on $\mathbb{T}$. The confusion with the earlier abstract symbols e_ρ is intentional: The functions e_ρ satisfy the law $e_\rho \cdot e_\sigma = e_{\rho+\sigma}$, with pointwise multiplication on the left, and the assignment "symbol e_ρ $\to$ function e_ρ" sets up an isomorphism of $\mathcal{I}$ with the character group (or Pontryagin dual) of $\mathbb{T}$, and also an isomorphism of the integral group ring $\mathbb{Z}\mathcal{I}$ of $\mathcal{I}$ with the ring $\mathcal{G}$ of ($\mathbb{C}$-valued continuous) functions on $\mathbb{T}$ generated by the characters e_ρ (the *character ring* or *representation ring* of $\mathbb{T}$). [One needs to know the - easily proved - fact that the e_ρ are linearly independent as functions on $\mathbb{T}$.] The algebraic structure of $\mathcal{G} \approx \mathbb{Z}\mathcal{I}$ is described by the formula $\mathbb{Z}[e_{\lambda_1}, (e_{\lambda_1})^{-1}, e_{\lambda_2}, (e_{\lambda_2})^{-2}, \ldots, e_{\lambda_l}, (e_{\lambda_l})^{-l}]$. It is fairly obvious, either from this structure or from the interpretation as functions on $\mathbb{T}$ that $\mathcal{G}$ is an integral domain (has no zero-divisors).

The two definitions for χ above, as element of $\mathbb{Z}\mathcal{I}$ or as element of $\mathcal{G}$, agree of course under the isomorphism of the two rings. In both cases we have $\chi = \Sigma m_\rho e_\rho$. As noted in the beginning, our aim is Weyl's formula for χ, and its consequences.

To begin with, the Weyl group $\mathcal{W}$ acts on $\mathfrak{h}_0^\top$ and on $\mathcal{I}$, and thus also (as ring automorphisms) on $\mathbb{Z}\mathcal{I}$; the formula is $Se_\rho = e_{S\rho}$. [In the function picture, i.e. for G, this means $Se_\rho(H) = e_\rho(S^{-1}H)$.] An element $a = \Sigma a_\rho e_\rho$ of $\mathcal{G}$ is called *symmetric* if Sa equals a for all S in $\mathcal{W}$, and *antisymmetric* or *skew* if Sa equals $\det S \cdot a$ for all S in $\mathcal{W}$. (Note that $\det S$ is 1 (resp -1) if S preserves (resp reverses) orientation of $\mathfrak{h}_0$.) The symmetric elements form a subring of $\mathcal{G}$; the product of a symmetric and a skew element is skew. It is important that the character of any rep is symmetric, by Theorem B (d) of §3.2.

For any ρ in $\mathcal{I}$ the sum of the elements of the orbit $\mathcal{W} \cdot e_\rho$ is a symmetric element. Just as easy and more important is the construction of skew elements: for ρ in $\mathcal{I}$ we put $A_\rho = \Sigma_\mathcal{W} \det S \cdot Se_\rho = \Sigma_\mathcal{W} \det S \cdot e_{S\rho}$. (The expression $\Sigma_\mathcal{W} \det S \cdot S$, an element of the integral group ring of $\mathcal{W}$, is called the *alternation operator*.) The element A_ρ is skew: For any T in $\mathcal{W}$ we have $TA_\rho = \Sigma \det S \cdot e_{TS\rho} = \det T \cdot \Sigma \det TS \cdot e_{TS\rho} = \det T \cdot \Sigma \det S \cdot e_{S\rho} = \det T \cdot A_\rho$ (we used the standard fact that TS runs once over $\mathcal{W}$ if S does). Note also $T \cdot A_\rho = A_{T\rho}$ by $T \cdot A_\rho = \Sigma \det S \cdot TSe_\rho = \Sigma \det S \cdot TST^{-1} \cdot Te_\rho = \Sigma \det S \cdot Se_{T\rho} = A_{T\rho}$ (we used $\det S = \det TST^{-1}$ and the fact that TST^{-1} also runs once over $\mathcal{W}$ if S does so).

PROPOSITION A.

(a) The element A_ρ is 0 if ρ is singular, i.e., lies on the infinitesimal Cartan-Stiefel diagram D' (of $\mathfrak{h}_0^\top$);

(b) For the other ρ there is exactly one e_σ from each Weyl chamber in A_ρ, with coefficient ± 1.

Part (b) is immediate from the definition of A_ρ. For (a) suppose we have $\langle \rho, \alpha \rangle = 0$ for some root α. Then $S_\alpha \rho$ equals ρ, and so $A_\rho = A_{S\rho} = S_\alpha \cdot A_\rho = -A_\rho$. √

Proposition A implies easily that the A_ρ with ρ strongly dominant (i.e., in $\mathcal{I}^0$) constitute a basis for the (free Abelian) group of skew elements of $\mathcal{G}$ (a sub group of the additive group of $\mathcal{G}$); in other words, that the skew elements are the finite sums $\Sigma a_\rho A_\rho$ with ρ in $\mathcal{I}^0$ and (unique) integers a_ρ.

We recall the element δ of $\mathcal{I}^0$, the sum of the fundamental weights λ_i. The associated element A_δ plays a special role. It happens that it factors very neatly, in several, equivalent, ways.

PROPOSITION B. $A_\delta = e_\delta \cdot \Pi_{\alpha>0}(1 - e_{-\alpha}) = e_{-\delta} \cdot \Pi_{\alpha>0}(e_\alpha - 1) = \Pi_{\alpha>0}(e_{\alpha/2} - e_{-\alpha/2})$. *(All products go over the positive roots.)*

The third product has to be understood properly. The terms $e_{\alpha/2}$ and $e_{-\alpha/2}$ do not make sense as elements of $\mathcal{G}$ (i.e., as functions on $\mathbb{T} = \mathfrak{h}_0/\mathcal{T}$), but they do make sense as functions (exponentials) on $\mathfrak{h}_0$ (or, if one wants, on the torus $\mathfrak{h}_0/2\mathcal{T}$; equivalently one could consider the integral group ring of the lattice $1/2\mathcal{I}$ or adjoin suitable square roots algebraically).

That the three products are equal comes from the fact that δ is one half the sum of all positive roots (Proposition B, §3.1); notice $e_{\alpha/2} - e_{-\alpha/2} = e_{\alpha/2}(1 - e_{-\alpha}) = e_{-\alpha/2}(e_\alpha - 1)$. We must show that they equal A_δ. The third product is antisymmetric, as follows from the formula $\det S = (-1)^r$, where $r = r_S$ is the number of positive roots sent to negative ones by S (§2.11, Corollary F); it is thus an integral linear combination of terms A_ρ with ρ in $\mathcal{I}^0$. Multiplying out the first product and collecting terms we see that e_δ appears with coefficient 1, since all other terms correspond to weights of the form $\delta - \Sigma\alpha$ with positive α's, which are lower than and different from δ. It is also clear that there is no other term than e_δ itself that comes from $\mathcal{I}^0$, since δ is already the lowest element of $\mathcal{I}^0$. But a sum of A_ρ's that has exactly the term e_δ coming form $\mathcal{I}^0$ must of course be just A_δ. $\checkmark$

Let now λ be a dominant weight and let φ_λ or just φ be the irrep (unique up to equivalence) with λ as extreme weight, operating on the vector space V; denote its character by χ_λ. We can now finally state *Weyl's character formula*, an important formula with many consequences [22].

THEOREM C. $\chi_\lambda = A_{\lambda+\delta}/A_\delta$

Note that the right-hand side is easy to write down (if one knows the Weyl group and δ), that it is fairly simple (except for being a quotient), and that one needs to know only the extreme weight λ, not the representation φ_λ.

The formula holds in the group ring $\mathcal{G}$. It says that $A_{\lambda+\delta}$ is divisible in $\mathcal{G}$ by A_δ and that the result is χ_λ. Another way to say this is that the relation $\chi_\lambda \cdot A_\delta = A_{\lambda+\delta}$ holds in $\mathcal{G}$; it determines χ_λ uniquely, in terms of $A_{\lambda+\delta}$ and A_δ, since $\mathcal{G}$ has no zero-divisors. One can also interpret the three terms of the formula as functions on $\mathbb{T}$ or $\mathfrak{h}_0$. There is some difficulty of course, since the denominator A_δ has lots of zeros. One can either rewrite the formula again as $\chi_\lambda \cdot A_\delta = A_{\lambda+\delta}$, or take the point of view that the function given by the quotient on the set where A_δ is not 0 extends, because of some miraculous cancelation of zeros, to the whole space, and the extended function is χ_λ.

Before we enter into the fairly long proof, we describe a simple example, namely the representations D_s of A_1 (cf.§1.11). Here a Cartan sub Lie algebra $\mathfrak{h}$ is given by $((H))$. The linear function $rH \to r$ (for r in $\mathbb{R}$) on $\mathfrak{h}_0$ is the fundamental weight λ_1 and also the element δ; the function $rH \to 2r$ is the unique positive root. The Weyl group contains besides the identity only the reflection $rH \to -rH$. The weights of the irrep D_s are the elements $n\lambda_1$ with $n = 2s, 2s-2, \ldots, -2s$; that is a restatement of the fact that these values are the eigenvalues of H in D_s. (In particular, the extreme weight λ for D_s is $2s\lambda_1$.) The character of D_s is then given by

$$\chi_s(rH) = \exp(2\pi i \cdot 2sr) + \exp(2\pi i \cdot (2s-2)r) + \cdots + exs(2\pi i \cdot -2sr)\ ;$$

writing $\exp(2\pi i r) = a$, this is the geometric series $a^{2s} + a^{(2s-2)} + \cdots + a^{-2s}$. On the other side we have

$$A_{\lambda+\delta} = \exp(2\pi i \cdot (2s+1)r) - \exp(2\pi i \cdot -(2s+1)r) = a^{2s+1} - a^{-(2s+1)}$$

and

$$A_\delta = \exp(2\pi i \cdot r) - \exp(2\pi i \cdot -r) = a - a^{-1}.$$

We see that Weyl's formula reduces to the usual formula for the geometric series.

We start on the proof of Theorem C. We shall interpret the elements of $\mathcal{G}$ as functions on $\mathfrak{h}_0$ (although in reality everything is completely formal, algebraic). For any given H_0 in $\mathfrak{h}_0$ we define the differentiation operator d_{H_0} (for $\mathbb{C}$-valued C^∞-functions on $\mathfrak{h}_0$) by

$$d_{H_0} f(H) = \lim_{t \to 0} (f(H + tH_0) - f(H))/2\pi i t\ .$$

All these operators commute, and one verifies $d_{H_0} e_\rho = \rho(H_0) e_\rho$ for any ρ in $\mathfrak{h}_0^\top$. Let $A_1, A_2, \ldots, A_l$ and $B_1, B_2, \ldots, B_l$ be any two dual bases of $\mathfrak{h}_0$ (and $\mathfrak{h}$) wr to the Killing form (so that $\langle A_i, B_j \rangle = \delta_{ij}$). We define the *Laplace operator* L as the sum $\Sigma d_{A_i} \circ d_{B_i}$ (this is independent of the choice of dual bases), and construct the bilinear operator ∇ by the relation $\mathrm{L}(fg) = \mathrm{L}f \cdot g + 2\nabla(f,g) + f \cdot \mathrm{L}g$. Explicitly we have $\nabla(f,g) = \Sigma d_{A_i} f \cdot d_{B_i} g + d_{B_i} f \cdot d_{A_i} g$. We note that ∇ is symmetric, vanishes if f or g is constant, and that it has the derivation property

$$\nabla(fg, h) = f \cdot \nabla(g,h) + \nabla(f,h) \cdot g.$$

Finally we have $\mathrm{L}e_\rho = \langle \rho, \rho \rangle e_\rho$ and $\nabla(e_\rho, e_\sigma) = \langle \rho, \sigma \rangle e_{\rho+\sigma}$ (this uses $\Sigma_i \rho(A_i) \cdot \rho(B_i) = \langle \rho, \rho \rangle$, which in turn comes from the duality of the bases $\{A_i\}$ and $\{B_i\}$).

We recall the root elements X_α, for α in Δ (§§2.4,2.5). We modify them to $x_\alpha = |\alpha|/\sqrt{2} X_\alpha$; the factors are chosen to have $\langle x_\alpha, x_{-\alpha} \rangle = 1$. Then $\{A_i, x_\alpha\}$ and $\{B_i, x_{-\alpha}\}$ are dual bases for $\mathfrak{g}$ wr to the Killing form. We define the *Casimir operator* Γ of φ as $\Sigma \varphi(A_i) \circ \varphi(B_i) + \Sigma_\Delta \varphi(x_\alpha) \circ \varphi(x_{-\alpha})$. (Again this is independent of any choices involved.)

This is not quite the Casimir operator of φ as defined in §3.4 (we are now using κ on $\mathfrak{g}$ and not t_φ on $\mathfrak{a}$); nevertheless the same computation shows that the new Γ commutes also with all $\varphi(X)$ for X in $\mathfrak{g}$ and is therefore, by Schur's lemma, a scalar operator $\gamma\, id$. (We show below that γ equals $\langle\lambda, \lambda\rangle + 2\langle\lambda, \delta\rangle$.)

In V we have the weight spaces V_ρ; we know the basic fact that $\varphi(x_\alpha)$ maps V_ρ into $V_{\rho+\alpha}$. Then $\varphi(x_\alpha) \circ \varphi(x_{-\alpha})$ and $\varphi(x_{-\alpha}) \circ \varphi(x_\alpha)$ both map V_ρ into itself; we write $t_{\alpha,\rho}$ and $t'_{\alpha,\rho}$ for the corresponding traces. There are two relations that are important for the proof of Weyl's formula.

LEMMA D.
(a) $m_\rho \cdot \langle\alpha, \rho\rangle = t_{\alpha,\rho} - t_{a,\rho+\alpha}$ *for all weights* ρ *of* φ *and all roots* α.

(b) $\Sigma_\alpha t_{\alpha,\rho} + m_\rho \cdot \langle\rho, \rho\rangle = m_\rho \cdot \gamma$ *for each weight* ρ.

(The sum in (b) goes over Δ, and γ is the eigenvalue of Γ described above.)

Proof: (a) The symmetry relation $\operatorname{tr} AB = \operatorname{tr} BA$ holds for any two linear transformations A and B that go in opposite directions between two vector spaces. Applied to $\varphi(x_\alpha)$ and $\varphi(x_{-\alpha)}$ on V_ρ and $V_{\rho+\alpha}$ this yields $t'_{\alpha,\rho} = t_{\alpha,\rho+\alpha}$. The relation $[X_\alpha X_{-\alpha}] = H_\alpha$ gives $[\varphi(x_\alpha), \varphi(x_{-\alpha})] = \langle\alpha, \alpha\rangle/2 \cdot \varphi(H_\alpha)$; on V_ρ this operator is scalar with eigenvalue $\langle\alpha, \alpha\rangle/2 \cdot \rho(H_\alpha) = \langle\alpha, \rho\rangle$. Taking the trace on V_ρ gives the result.

(b) On V_ρ the eigenvalue of $\varphi(A_i) \circ \varphi(B_i)$ is $\rho(A_i) \cdot \rho(B_i)$; the sum of these values is $\langle\rho, \rho\rangle$. Taking the trace of Γ on V_ρ gives the result. $\surd$

The next lemma contains the central computation.

LEMMA E. $A_\delta \cdot \chi_\lambda$ *is eigen element of the Laplace operator* L, *with eigenvalue* $\langle\delta, \delta\rangle + \gamma$.

Proof: We have $\mathrm{L}(A_\delta \cdot \chi_\lambda) = \mathrm{L}A_\delta \cdot \chi_\lambda + 2\nabla(A_\delta, \chi_\lambda) + A_\delta \cdot \mathrm{L}\chi_\lambda$. The properties of L listed above and the invariance of $\langle,\rangle$ under the Weyl group imply the relation $\mathrm{L}A_\delta = \langle\delta, \delta\rangle A_\delta$. From $\chi_\lambda = \Sigma m_\rho e_\rho$ we get $\mathrm{L}\chi_\lambda = \Sigma m_\rho \langle\rho, \rho\rangle e_\rho$. Substituting for $m_\rho \cdot \langle\rho, \rho\rangle$ from Lemma D,(b), we obtain:

$$\mathrm{L}(A_\delta \cdot \chi_\lambda) = (\langle\delta, \delta\rangle + \gamma) A_\delta \cdot \chi_\lambda + (2\nabla(A_\delta . \chi_\lambda) - \Sigma_{\rho,\alpha} t_{\alpha,\rho} A_\delta \cdot e_\rho).$$

We show now that the second term is 0, after multiplying it by A_δ (this will establish Lemma E, since there are no zero divisors in $\mathcal{G}$); from Proposition B we have $A_\delta^2 = \epsilon \cdot \Pi_\beta(e_\beta - 1)$ with $\epsilon = \pm 1$. We use the properties of ∇, in particular the derivation property, repeatedly. We have

$$\begin{aligned} 2A_\delta \cdot \nabla(A_\delta, \chi_\lambda) &= \epsilon \cdot \nabla(\Pi_\beta(e_\beta - 1), \chi_\lambda) \\ &= \epsilon \cdot \Sigma_\rho m_\rho \nabla(\Pi_\beta(e_\beta - 1), e_\rho) \\ &= \epsilon \cdot \Sigma_\rho m_\rho \Sigma_\alpha \Pi_{\beta \neq \alpha}(e_\beta - 1)\langle \alpha, \rho \rangle e_{\alpha+\rho} \end{aligned}$$

By (a) of Lemma D, this equals $\epsilon \cdot \Sigma_{\rho,\alpha}\Pi_{\beta\neq\alpha}(e_\beta - 1)(t_{\alpha,\rho} - t_{\alpha,\rho+\alpha})e_{\alpha+\rho}$. Replacing $\rho + \alpha$ by ρ in the terms of the sum involving $t_{\alpha,\rho+\alpha}e_{\alpha+\rho}$ we get $\epsilon \cdot \Sigma_{\rho,\alpha}\Pi_{\beta\neq\alpha}((e_\beta - 1)t_{\alpha,\rho}(e_{\alpha+\rho} - e_\rho)$. With $e_{\alpha+\rho} - e_\rho = (e_\alpha - 1)e_\rho$ this turns into $\epsilon \cdot \Sigma_{\rho,\alpha}\Pi_\beta(e_\beta - 1)t_{\alpha,\rho}e_\rho$, which is the same as $A_\delta \cdot \Sigma_{\rho,\alpha}t_{\alpha,\rho}A_\delta \cdot e_\rho$. $\surd$

We come to the proof of Weyl's formula. By Lemma E all the terms e_ρ appearing in $A_\delta \cdot \chi_\lambda$ have the same $\langle \rho, \rho \rangle$, $= \langle \delta, \delta \rangle + \gamma$. If we multiply the expressions $A_\delta = \Sigma \det S \cdot e_{S\delta}$ and $\chi_\lambda = \Sigma m_\rho e_\rho = e_\lambda + \cdots$, we get a sum of terms of the form $r \cdot e_{S\delta+\rho}$ with integral r. The term $e_{\delta+\lambda}$ appears with coefficient 1, since δ is maximal among the $S\delta$ (by Proposition B of §3.1) and λ is maximal among the ρ in χ_λ (by Corollary D in §2). Thus all the terms for which $\langle S\delta + \rho, S\delta + \rho \rangle$ is different from $\langle \delta + \lambda, \delta + \lambda \rangle$ must cancel out. (We see also that $\langle \delta, \delta \rangle + \gamma$ equals $\langle \delta + \lambda, \delta + \lambda \rangle$, so that we have $\gamma = \langle \lambda, \lambda \rangle + 2\langle \lambda, \delta \rangle$.)

Suppose now that $\langle S\delta + \rho, S\delta + \rho \rangle$ equals $\langle \delta + \lambda, \delta + \lambda \rangle$. Then we have also $\langle \delta + S^{-1}\rho, \delta + S^{-1}\rho \rangle = \langle \delta + \lambda, \delta + \lambda \rangle$. Here $S^{-1}\rho$, $= \sigma$ say, is also a weight in χ_λ. We show that for any such σ, except for λ itself, the norm square $\langle \delta + \sigma, \delta + \sigma \rangle$ is strictly less than $\langle \delta + \lambda, \delta + \lambda \rangle$, as follows. We know that (a) $\langle \lambda, \lambda \rangle$ is maximal among the $\langle \sigma, \sigma \rangle$, that (b) $\lambda - \sigma$ is a linear combination of the fundamental roots α_i with non-negative integral coefficients, and that (c) $\langle \delta, \alpha_i \rangle$ is positive (since $\delta(H_i) = 1$). Thus $\langle \delta, \delta \rangle + 2\langle \delta, \sigma \rangle + \langle \sigma, \sigma \rangle < \langle \delta, \delta \rangle + 2\langle \delta, \lambda \rangle + \langle \lambda, \lambda \rangle$. So the σ above is λ, and the ρ above is $S\lambda$. This means that $A_\delta \cdot \chi_\lambda$ contains only of the form $r \cdot e_{S(\lambda+\delta)}$, i.e. only terms that (up to an integral factor) appear in $A_{\lambda+\delta}$; since it is a skew element and contains $e_{\lambda+\delta}$ with coefficient 1, it clearly must equal $A_{\lambda+\delta}$. $\surd$

3.8 Some consequences of the character formula

The first topic is *Weyl's degree formula* for the irrep φ_λ with extreme weight λ; it gives the dimension of the vector space in which the representation takes place.

THEOREM A. *The degree d_λ of φ_λ is $\Pi_{\alpha>0}\langle \alpha, \lambda + \delta \rangle / \Pi_{\alpha>0}\langle \alpha, \delta \rangle$. (The products go over all positive roots.)*

(This could also be written $d_\lambda = \Pi_{\alpha>0}(\lambda + \delta)(H_\alpha)/\Pi_{\alpha>0}\delta(H_\alpha)$.)

Proof: The degree in question is the value of χ_λ at $H = 0$. Unfortunately both $A_{\lambda+\delta}$ and A_δ have zeros of high order at 0. Thus we must take derivatives before we can substitute 0 (L'Hopital). We use the root vectors h_α (see

§2.4), and apply the differential operator $d = \Pi_{\alpha>0} d_{h_\alpha}$ to both sides of the equation $A_\delta \cdot \chi_\lambda = A_{\lambda+\delta}$. Using the factorization $A_\delta = e_{-\delta} \cdot \Pi_{\alpha>0}(e_\alpha - 1)$ and differentiating out (Leibnitz's rule) one sees that for $H = 0$ the relation $d(A_\delta \cdot \chi_\lambda)(0) = dA_\delta(0) \cdot \chi_\lambda(0)$ holds. Thus d_λ is the quotient of $dA_{\lambda+\delta}(0)$ and $dA_\delta(0)$. From $d_{h_\alpha} e_\rho = \rho(h_\alpha) e_\rho = \langle \rho, \alpha \rangle e_\rho$ we find $de_\delta(0) = \Pi_{\alpha>0}\langle \delta, \alpha \rangle$. Similarly we get $de_{S\delta}(0) = \Pi_{\alpha>0}\langle S\delta, \alpha\rangle = \Pi_{\alpha>0}\langle \delta, S^{-1}\alpha\rangle$. Now some of the $S^{-1}\alpha$ are negative roots; from §2.11, Corollary F we see that the last product is exactly $\det S \cdot \Pi_{\alpha>0}\langle \delta, \alpha\rangle$. Thus all terms in $A_\delta = \Sigma \det S \cdot e_{S\delta}$ contribute the same amount, and so $dA_\delta(0)$ equals $|\mathcal{W}| \cdot \Pi_{\alpha>0}\langle \alpha, \delta\rangle$. The corresponding result for $A_{\lambda+\delta}$ finishes the argument. $\surd$

Our next topic is Kostant's formula for the multiplicities of the weights [17]. One defines the *partition function* $\mathcal{P}$ on the set $\mathcal{I}$ (lattice of weights) by:

$\mathcal{P}(\rho)$ is the number of (unordered) partitions of ρ into positive roots; in detail this is the number of systems $(p_\alpha)_{\alpha>0}$ of non-negative integers p_α satisfying $\rho = \Sigma_{\Delta^+} p_\alpha \alpha$. Note that $\mathcal{P}(\rho)$ is 0 for many ρ, in particular for every non-positive weight except 0 and for any weight not in the root lattice $\mathcal{R}$.

We continue with the earlier notation; λ a dominant weight, φ or φ_λ the irrep with extreme weight λ, and $\chi_\lambda = \Sigma m_\rho e_\rho$ the character of φ_λ.

PROPOSITION B (KOSTANT'S FORMULA). *The multiplicity m_ρ of ρ in φ_λ is $\Sigma_{S\mathcal{W}} \det S \cdot \mathcal{P}(S(\lambda+\delta) - \delta - \rho)$.*

This rests on the formal relation $(\Pi_{\alpha>0}(1 - e_{-\alpha}))^{-1} = \Sigma calP(\rho) e_{-\rho}$, obtained by multiplying together the expansions $1/(1 - e_{-\alpha}) = 1 + e_{-\alpha} + e_{-2\alpha} + \cdots$. To make sense of these formal infinite series, we let E stand for the cone in $\mathfrak{h}_0^\top$ spanned by the positive roots with non-positive coefficients (the *backward cone*); for any μ in $\mathfrak{h}_0^\top$ we set $E_\mu = \mu + E$. Now we extend the group ring $\mathcal{G}$ (finite integral combinations of the e_ρ) to the ring $\mathcal{G}^\infty$ consisting of those formal infinite series $\Sigma c_\rho e_\rho$ (with integral c_ρ), whose support (the set of ρ's with non-zero c_ρ) is contained in some E_μ; the restriction on the support is analogous to considering power series with a finite number of negative exponents and makes it possible to not only add, but also multiply these elements in the obvious way (using some simple facts about the cones E_μ). E.g., the series $\Sigma\mathcal{P}(\rho)e_{-\rho}$ above has its support in $E_0 = E$.

With the help of Proposition B in §3.6 we write Weyl's formula in the form $\chi_\lambda = A_{\lambda+\delta} e_{-\delta} / \Pi_{\alpha>0}(1 - e_{-\alpha})$, which with our expansion of the denominator becomes $\Sigma m_\rho e_\rho = (\Sigma \det S \cdot e_{S(\lambda+\delta)-\delta}) \cdot (\Sigma \mathcal{P}(\sigma) e_{-\sigma})$. Multiplying out, we see that we get m_ρ, for a given ρ, by using those σ for which $S(\lambda+\delta) - \delta - \sigma$ equals ρ for some S in $\mathcal{W}$. That is just what Kostant's formula says. $\surd$

While the formula is very explicit, it is also very non-computable, to a minor extent because of the summation over the Weyl group, but mainly because of the difficulty of evaluating the partition function (cf. the case of partitions of the natural numbers!). We present two more practical algorithms for the computation of the multiplicities of the weights of φ_λ.

The first one is Klimyk's formula [15].

For any integral linear form ρ we put $\epsilon_\rho = \det T$, if there exists T in the Weyl group with $\rho = T(\lambda+\delta)-\delta$, and $=0$ otherwise. (The operations $\lambda \to T(\lambda+\delta)-\delta$ constitute the *shifted* action of $\mathcal{W}$ on $\mathfrak{h}_0^{\top}$, with $-\delta$ as origin.)

PROPOSITION C (KLIMYK'S FORMULA). *For any weight ρ in $\mathcal{I}$ the multiplicity m_ρ equals $\epsilon_\rho - \Sigma_{S\neq \mathrm{id}} \det S \cdot m_{\rho+\delta-S\delta}$.*

We first comment on the formula and then prove it.

The main point is that that $\delta - S\delta$, for $S \neq \mathrm{id}$, is a non-zero sum of positive roots (see Proposition B of §3.1). Thus m_ρ is expressed as a sum of a fixed number (namely $|\mathcal{W}|-1$ terms) of the multiplicities of weights that are higher than ρ (the $\rho+\delta-S\delta$), plus the term ϵ_ρ (which requires a check over the Weyl group). Thus we get an inductive (wr to the order in $\mathfrak{h}_0^{\top}$) computation of m_ρ. It begins with $m_\lambda = 1$. This is quite practical, particularly of course for cases of low rank and small Weyl group. The main objection to the formula is that about half the terms are negative, because of the factor $\det S$, and that therefore there will be a lot of cancelation to get the actual values. (The next approach, Freudenthal's formula, avoids this.)

Now the proof: We rewrite Weyl's formula as $\Sigma m_\rho e_\rho \cdot \Sigma \det S \cdot e_{S\delta} = \Sigma \det T \cdot e_{T(\lambda+\delta)}$. The left-hand side can be written first as $\Sigma_{\mathcal{W}}(\Sigma_{\mathcal{I}} \det S \cdot m_\rho e_\rho e_{S\delta})$, then as $\Sigma_{\mathcal{W}}(\Sigma_{\mathcal{I}} \det S \cdot m_{S\rho} e_{S(\rho+\delta)})$ (since $S\rho$, for fixed S, runs over $\mathcal{I}$ just as ρ does), and then (putting $S(\rho+\delta) = \sigma+\delta$ with σ, for fixed S, again running once over $\mathcal{I}$) as $\Sigma_{\mathcal{W}}(\Sigma_{\mathcal{I}} \det S \cdot m_{\sigma+\delta-S\delta} e_{\sigma+\delta})$, which equals $\Sigma_{\mathcal{I}}(\Sigma_{\mathcal{W}} \det S \cdot m_{\rho+\delta-S\delta}) e_{\rho+\delta}$. Comparing this with the right-hand side of Weyl's formula, we see that the coefficient of $e_{\rho+\delta}$ is $\det T$, if $\rho+\delta = T(\lambda+\delta)$ for some (unique) T in $\mathcal{W}$, and 0 otherwise. That's just what Klimyk's formula says. $\surd$

We now come to Freudenthal's formula [8].

PROPOSITION D. *The multiplicities m_ρ satisfy the relation*

$$(\langle\lambda+\delta,\lambda+\delta\rangle - \langle\rho+\delta,\rho+\delta\rangle)\cdot m_\rho = 2\Sigma_{\alpha>0}\Sigma_1^\infty m_{\rho+t\alpha}\langle\rho+t\alpha,\alpha\rangle.$$

We first comment on the formula and then prove it.

We saw at the end of §3.6 that for any weight ρ of φ (i.e., with $m_\rho \neq 0$) the inequality $\langle\lambda+\delta,\lambda+\delta\rangle - \langle\rho+\delta,\rho+\delta\rangle > 0$ holds. Thus the formula gives m_ρ inductively, in terms of the multiplicities of the strictly greater weights $m_{\rho+t\alpha}$ for $t \geq 1$ and α in Δ^+. The "induction" again begins with $\rho = \lambda$. All the terms in the formula are non-negative, so there is no such cancelation as in Klimyk's formula, which makes the formula quite practical. On the other hand, in contrast to Klimyk's formula, the number of terms in the sum on the right is not fixed, and becomes larger and larger, for some ρ's, as λ gets larger.

Now to the proof:

First we state some results of A_1-representation theory (§1.11) in a slightly different form; we use the notation developed there.

Lemma E.

(a) In any irrep D_s of A_1 the sum of the eigenvalues of H on all the v_j (the trace of H) is 0;

(b) Each vector v_i is eigenvector of the operator X_+X_-; the eigenvalue equals the sum of the eigenvalues of H on the vectors $v_0, v_1, \ldots, v_i$.

We return to our irrep φ_λ, on the vector space V, with extreme weight λ, weight spaces V_ρ, etc., as in the last few sections.

We choose a root α (positive or negative) and consider the sub Lie algebra $\mathfrak{g}^{(\alpha)} = ((H_\alpha, X_\alpha, X_{-\alpha}))$ of type A_1 (see §2.5).

Lemma F. *There exists a decomposition of V under the action of $\mathfrak{g}^{(\alpha)}$ into irreducible subspaces $W_u, u = 1, 2, 3, \ldots$ (each equivalent to some standard rep D_s) such that the eigenvectors of H_α in any W_u are weight vectors of φ.*

Proof: For a given weight ρ we form its α-string, the direct sum of the weight spaces $V_{\rho+t\alpha}$ with $t \in \mathbb{Z}$. By the basic lemma A of §2 it is $\mathfrak{g}^{(\alpha)}$-invariant. Any decomposition into irreducible subspaces clearly has the property described in Lemma F. And V is direct sum of such strings. $\surd$

Note that the eigenvalues of H_α in any W_u are of the form $(\rho + t\alpha)(H_\alpha)$ for some ρ and some t-interval $a \leq t \leq b$; also recall $\alpha(H_\alpha) = 2$, consistent with the nature of the D_s's. $\surd$

The next lemma is one of the main steps to Freudenthal's formula. We recall the elements $x_\alpha = |\alpha|/\sqrt{2}X_\alpha$ introduced in §6.

Lemma G.

(a) For any integral form ρ the sum $\Sigma_{-\infty}^{\infty} m_{\rho+t\alpha}\langle\rho + t\alpha, \alpha\rangle$ is 0.

(b) For any such ρ the trace of the operator $x_\alpha x_{-\alpha}$ on the weightspace V_ρ is $\Sigma_0^\infty m_{\rho+t\alpha}\langle\rho + t\alpha, \alpha\rangle$.

[If we use $X_\alpha X_{-\alpha}$ instead of $x_\alpha x_{-\alpha}$, then $\langle\rho + t\alpha, \alpha\rangle(= (\rho + t\alpha)(h_\alpha))$ becomes $(\rho + t\alpha)(H_\alpha)$.]

Part (a) is an immediate consequence of Lemma E (a), applied to the decomposition of V into $\mathfrak{g}^{(\alpha)}$-irreducible subspaces described in Lemma F. We are in effect summing all the eigenvalues of H_α in all those W_u that intersect some $V_{\rho+t\alpha}$ non-trivially; for each W_u we get 0.

Part (b) follows similarly. This time we consider only those W_u that meet some $V_{\rho+t\alpha}$ with $t \geq 0$ non-trivially. The right-hand side consists of two parts: (1) The sum over the W_u that meet V_ρ itself non-trivially. This gives the trace of $x_\alpha x_{-\alpha}$ on V_ρ, by (b) of Lemma E, since for each W_u we are summing the eigenvalues of H_α from $\rho(H_\alpha)$ on up. (2) The sum over the W_u that don't meet V_ρ, but meet $V_{\rho+t\alpha}$ for some positive t. This gives 0 by (a) of lemma E as before; for each W_u we are summing all the eigenvalues of H_α. √

The next ingredient is the Casimir operator Γ, introduced in §3.6. We saw there that Γ acts as the scalar operator $\langle\lambda + 2\delta, \lambda\rangle \mathrm{id}_V$. Thus the trace of Γ on V_ρ is $m_\rho\langle\lambda + 2\delta, \lambda\rangle$. On the other hand, from the definition of Γ we get two parts for this trace, one corresponding to $\mathfrak{h}$ and one corresponding to the roots. The first part yields $m_\rho \Sigma \rho(A_i)\rho(B_i)$, which equals $m_\rho\langle\rho, \rho\rangle$. The second gives $\Sigma_\Delta \Sigma_0^\infty m_{\rho+t\alpha}\langle\rho + t\alpha, \alpha\rangle$, by Lemma G (a). Here we can start the sum at $t = 1$ instead of $t = 0$, since for $t = 0$ the contributions of each pair $\{\alpha, -\alpha\}$ of roots cancel. We now divide Δ into Δ^+ and Δ^-, and note that by Lemma G (b) we have $\Sigma_1^\infty m_{\rho+t\alpha}\langle\rho + t\alpha, \alpha\rangle = -m_\rho\langle\rho, \alpha\rangle - \Sigma_{-1}^{-\infty} m_{\rho+t\alpha}\langle\rho + t\alpha, \alpha\rangle$, for any α. Taking α in Δ^-, we can rewrite this as $m_\rho\langle\rho, -\alpha\rangle + \Sigma_1^\infty m_{\rho+t\cdot-\alpha}\langle\rho + t\cdot -\alpha, -\alpha\rangle$. Thus the value of the second sum in Γ becomes $\Sigma_{\alpha>0} m_\rho\langle\rho, \alpha\rangle + 2\Sigma_{\alpha>0} m_{\rho+t\alpha}\langle\rho + t\alpha, \alpha\rangle$; here the first term equals $2m_\rho\langle\rho, \delta\rangle$. All in all we get for the trace of Γ on V_ρ the value $m_\rho\langle\rho + 2\delta, \rho\rangle + 2\Sigma_{\alpha>0}\Sigma_1^\infty m_{\rho+t\alpha}\langle\rho + t\alpha, \alpha\rangle$. (Incidentally, from this computation we get once more the eigenvalue γ of Γ: For $\rho = \lambda$ we have $m_\lambda = 1$, as we know, and the sum vanishes, since all the $m_{\rho+t\alpha}$ are 0.) Equating the two values for this trace we get Freudenthal's formula. √

Our last topic is the generalization of the Clebsch-Gordan series, i.e. the problem of decomposing the tensor product of two irreps. Let λ' and λ'' be two dominant weights, with the corresponding irreps φ' and φ''. By complete reducibility the tensor product $\varphi' \otimes \varphi''$ splits as $\Sigma n_\lambda \varphi_\lambda$ (the sum goes over $\mathcal{I}^d$, the dominant weights, and is finite of course) with *multiplicities* n_λ. The problem is to determine the n_λ. (We know already from the discussion of the Cartan product that $\lambda' + \lambda''$ is the highest of the λ occurring here, and that $n_{\lambda'+\lambda''}$ is 1.) We put $n_\lambda = 0$ for any non-dominant λ.

We consider three approaches: Steinberg's formula [22], R. Brauers's algorithm [2], and Klimyk's formula [15].

PROPOSITION H (STEINBERG'S FORMULA)
$n_\lambda = \Sigma_{S,T\in\mathcal{W}} \det ST \cdot \mathcal{P}(S(\lambda' + \delta) + T(\lambda'' + \delta) - \lambda - 2\delta)$ *for any dominant* λ.

(Here $\mathcal{P}$ is the partition function, described above.)

The formula is very explicit, but not very practical: There is a double summation over the Weyl group, and the partition function, difficult to evaluate, is involved.

For the proof we write m'_ρ for the multiplicities of the weights of λ'. It is clear that the character of $\varphi' \otimes \varphi''$ is the product of the characters of φ' and φ'' (tensor products of weight vectors are weight vectors with the sum weight); thus $\chi_{\lambda'}\chi_{\lambda''} = \Sigma n_\lambda \chi_\lambda$. Using Weyl's formula we rewrite this as $\chi_{\lambda'} \cdot A_{\lambda''+\delta} = \Sigma n_\lambda A_{\lambda+\delta}$. Applying Kostant's formula for the m'_ρ we get

$$\Sigma_\rho(\Sigma_S \det S \cdot \mathcal{P}(S(\lambda' + \delta) - \rho - \delta) \cdot e_\rho \cdot A_{\lambda''+\delta} = \Sigma_\lambda n_\lambda A_{\lambda+\delta}.$$

On multiplying out this becomes

$$\Sigma_{\rho,S,T} \det ST \cdot \mathcal{P}(S(\lambda' + \delta) - \rho - \delta) \cdot e_{T(\lambda''+\delta)+\rho} = \Sigma_{\lambda,S} n_\lambda \cdot \det S \cdot e_{S(\lambda+\delta)}.$$

We have to collect terms and compare coefficients.

On the right we change variables, putting, for fixed S, $S(\lambda + \delta) = \sigma + \delta$, and obtaining $\Sigma_{\sigma,S} \det S \cdot n_{S^{-1}(\sigma+\delta)-\delta} \cdot e_{\sigma+\delta}$ or $\Sigma_\sigma \Sigma_S \det S \cdot n_{S(\sigma+\delta)-\delta} \cdot e_{\sigma+\delta}$. On the left we put, for fixed S and T, $T(\lambda'' + \delta) + \rho = \sigma + \rho$ and obtain $\Sigma_{\sigma,S,T} \det ST \cdot \mathcal{P}(S(\lambda' + \delta) + T(\lambda'' + \delta) - \sigma - 2\delta) \cdot e_{\sigma+\delta}$. Finally we note: If σ is dominant, then $\sigma + \delta$ is strongly dominant, and then for $S \neq id$ the weight $S(\sigma + \delta) - \delta$ is not dominant and so $n_{S(\lambda+\delta)-\delta}$ is 0. Thus for dominant σ the coefficient of $e_{\sigma+\delta}$ on the expression for the right is n_σ, and the coefficient of $e_{\sigma+\delta}$ in the expression for the left-hand side is the value stated in Steinberg's formula. $\surd$

Next we come to R. Brauer's algorithm. It is based on the assumption that the weights of one of the two representations φ' and φ'' are known, so that we have, say, $\chi_{\lambda'} = \Sigma m'_\rho e_\rho$. We write the decomposition relation $\chi_{\lambda'} \cdot \chi_{\lambda''} = \Sigma n_\lambda \chi_\lambda$, using Weyl's formula for $\chi_{\lambda''}$ and the χ_λ and multiplying by A_δ, in the form

$$(\Sigma m'_\rho e_\rho) \cdot A_{\lambda''+\delta} = \Sigma n_\lambda A_{\lambda+\delta}.$$

We see that the problem amounts to expressing the skew element on the left in terms of the standard skew elements $A_{\lambda+\delta}$ with dominant λ.

Brauer's idea is to relax this and to admit terms with non-dominant λ (each one of which is of course up to sign equal to one with a dominant λ). For an arbitrary weight τ we write $[\tau]$ for the unique dominant weight in the $\mathcal{W}$-orbit of τ (i.e., the element in $\mathcal{W}\cdot\tau$ that lies in the fundamental Weyl chamber), and we write $\eta_\tau = 0$, if τ is singular (lies in some singular plane $(\alpha, 0)$), and $=\det S$ where S, in the Weyl group, is the unique element with $S\tau = [\tau]$, for regular τ. We have then $A_\tau = \eta_\tau A_{[\tau]}$ for any τ. (Recall $A_\tau = 0$ for singular τ.)

PROPOSITION I (R. BRAUER'S ALGORITHM).
$\chi_{\lambda'} \cdot A_{\lambda''+\delta} = \Sigma m'_\sigma \cdot \eta_{\sigma+\lambda''+\delta} \cdot A_{[\sigma+\lambda''+\delta]}$, where the sum goes over the set of weights σ of φ'.

Proof: We introduce the set $E = \{(e_\rho, e_\mu) : \rho \text{ weight of } \varphi', \mu \in \mathcal{W}\cdot(\lambda''+\delta)\}$, the product of the set of weights of φ' and the Weyl group orbit of $\lambda''+\delta$. To each element (e_ρ, e_μ) of E, with $\mu = S(\lambda''+\delta)$, we assign the term $m'_\rho \det S\, e_\rho \cdot e_\mu$; the sum of all these terms is then precisely the left-hand side in Proposition I. Now we let $\mathcal{W}$ operate *diagonally* on E, with $S(e_\rho, e_\mu) = (Se_\rho, Se_\mu)$. Each orbit contains an element of the form $(e_\sigma, e_{\lambda''+\delta})$ (and different e_σ's correspond to different orbits). Using the invariance of the weights under $\mathcal{W}$ (i.e. $m'_{S\sigma} = m'_\sigma$) we see that the sum of the terms corresponding to this orbit, i.e., $\Sigma_{\mathcal{W}} m'_{S\sigma}\cdot\det S\cdot e_{S\rho}\cdot e_{S(\lambda''+\delta)}$, is precisely $m'_\sigma A_{\sigma+\lambda''+\delta}$, equal to the term corresponding to σ on the right hand side of the formula. $\surd$

We restate the result in the form given by Klimyk.

PROPOSITION J. *For dominant λ the multiplicity n_λ equals $\Sigma m'_\sigma \cdot \eta_{\sigma+\lambda''+\delta}$, where the sum goes over those weights σ of φ' that satisfy $[\sigma + \lambda'' + \delta] = \lambda + \delta$.*

3.9 Examples

We start with some examples for the degree formula, §3.8, Theorem F. We shall work out the degrees of the spin representations Δ and Δ^+, Δ^- of B_l and D_l. With the conventions of §2.13, the Killing form agrees with the Pythagorean inner product up to a factor; by homogeneity of the degree formula we can suppress this factor. For B_l the positive roots are the ω_i and the $\omega_i \pm \omega_j$ with $i < j$; the lowest weight δ is $(l-1/2)\omega_1 + (l-3/2)\omega_2 + \cdots$; the extreme weight for the spin representation is $\lambda_l = 1/2(\omega_1+\omega_2+\cdots+\omega_l)$. With $\langle\omega_i, \delta\rangle = l-i+1/2$ and $\langle\omega_i, \lambda_l+\delta\rangle = l-i+1$ the formula evaluates to

$$\begin{aligned}
&\frac{\prod_i (l-i+1)\cdot\prod_{i<j}(j-i)(2l+2-i-j)}{\prod_i (l-i+1/2)\cdot\prod_{i<j}(j-i)(2l+1-i-j)}\\
&= \prod_{0\le i\le l-1}\frac{(2i+2}{2i+1)}\cdot\prod_{0\le i<j\le l-1}\frac{i+j+2}{i+j+1}\\
&= 2^l\cdot\prod_{0\le i\le l-1}\frac{i+1}{2i+1}\cdot\frac{\prod_{1\le i<j\le l-1} i+j+2}{\prod_{0\le i<j\le l-1} i+j+1}\\
&= 2^l\cdot\prod_{0\le i\le l-1}\frac{i+1}{2i+1}\cdot\frac{\prod_{1\le i\le l-1} 2i+1}{\prod_{0<j\le l-1} j+1}\\
&= 2^l
\end{aligned}$$

For D_l the positive roots are the $\omega_i \pm \omega_j$ with $i<j$; the lowest form δ is $(l-1)\omega_1+(l-2)\omega_2+\cdots$; the extreme weights for the two spin representations are $\lambda_{l-1}, \lambda_l = 1/2(\omega_1+\omega_2+\cdots\mp\omega_l)$. For φ_l the formula gives

$$\begin{aligned}
\text{degree } \varphi_l &= \frac{\prod_{i<j}(j-i)(2l-i-j+1)}{\prod_{i<j}(j-i)(2l-i-j)}\\
&= \prod_{0\le i<j\le l-1}\frac{i+j+1}{i+j}\\
&= \frac{\prod_{1\le i\le j\le l-1} i+j}{\prod_{0\le i<j\le l-1} i+j}\\
&= \frac{\prod_{1\le i\le l-1} 2i}{\prod_{0<j\le l-1} j}\\
&= 2^{l-1}
\end{aligned}$$

Making the appropriate modification for φ_{l-1} we get for its degree the value

$$\text{degree } \varphi_{l-1} = \text{degree } \varphi_l \cdot \prod_{i<l}\frac{(l-i+1)}{(l-i)}\cdot\frac{(l-i)}{(l-i+1)} = 2^{l-1}$$

A consequence of this computation is the following: The weights of Δ are exactly the $1/2(\pm\omega_1\pm\omega_2\pm\cdots\pm\omega_l)$ since all these must occur by invariance under the Weyl group and they are already $2l$ in number. Similarly for Δ^+ and Δ^- the weights are exactly the $1/2(\pm\omega_1\pm\omega_2\pm\cdots\pm\omega_l)$ with an even, respectively odd, number of minus signs.

In the same vein the weights of any Λ_r are the $\omega_{i_1}+\cdots+\omega_{i_p}-(\omega_{j_1}+\cdots+\omega_{j_q})$ with $i_1<\cdots<i_p, j_1<\cdots<j_q$, and $p+q=r-1$ or r for B_l and $=r$ for D_l. (The difference comes from the fact that 0 is a weight of Λ_1 for B_l, but not for D_l.)

Returning to the general degree formula we write λ as $\Sigma n_i\lambda_i$, in terms of the fundamental weights λ_i. Clearly the formula gives the degree of the rep φ_λ as a polynomial in the variables n_i, of degree $1/2(\dim\mathfrak{g} - \operatorname{rank}\mathfrak{g})$ (equal to the number of positive roots). It is fairly customary to write $\lambda + \delta = \Sigma g_i\omega_i$, thus expressing the degree as a polynomial in the g_i.

As an example for the various constructions we consider $A_2 = \mathfrak{sl}(3,\mathbb{C})$ in more detail. (The simply connected Lie group here is $SL(3,\mathbb{C})$. The corresponding compact group—which has the same representations as $SL(3,\mathbb{C})$ and $\mathfrak{sl}(3,\mathbb{C})$—is $SU(3)$; it is of interest in physics under the heading "the eightfold way". The point is that the elementary particles in nature appear to occur in families that correspond to the weight systems of the irreps of A_2. For instance, the two fundamental irreps φ_1 and φ_2, both of dimension three with three weights of multiplicities 1, correspond to the two systems of quarks and antiquarks. The adjoint rep, of dimension eight (hence the "eightfold way"), corresponds to a family of eight particles.)

To begin with, from the description of the roots (§2.14) we find for the degree d_λ, with $\lambda = n_1\lambda_1 + n_2\lambda_2 = g_1\omega_1 + g_2\omega_2$, the expression

$$\frac{\Pi\langle\omega_i - \omega_j, g_1\omega_1 + g_2\omega_2\rangle}{\Pi\langle\omega_i - \omega_j, 2\omega_1 + \omega_2\rangle},$$

with $g_1 > g_2 > 0$ and the product running over $1 \leq i < j \leq 3$. With $\langle\omega_i,\omega_j\rangle = \delta_{ij}$ this becomes $d_\lambda = 1/2 g_1 \cdot g_2 \cdot (g_1 - g_2)$ or $1/2(n_1+1)(n_2+1)(n_1+n_2+2)$.

For $g_1 = 2, g_2 = 1$ this is 1; the rep is the trivial one, φ_0. For $g_1 = 3, g_2 = 1$ the degree is 3, with $\lambda = \omega_1 (= \lambda_1)$; the rep is Λ_1, i.e., $\mathfrak{sl}(3,\mathbb{C})$ itself. For $g_1 = 3, g_2 = 2$ we get again 3, with $\lambda = \omega_1 + \omega_2 (= -\omega_3 = \lambda_2)$; this is Λ_2, the contragredient rep, i.e. the negative transpose. For $g_1 = 4, g_2 = 1$ we get 6 for the degree, with $\lambda = 2\lambda_1 = 2\omega_1$; this is the symmetric square of Λ_1, i.e., the rep on the quadratic polynomials on $\mathbb{C}^3$. For $g_1 = 4, g_2 = 3$ we get the contragredient, $\lambda = 2\lambda_2 = 2\omega_1 + 2\omega_2$. Finally, $g_1 = 4, g_2 = 2$ gives degree 8, with $\lambda = \lambda_1 + \lambda_2 = 2\omega_1 + \omega_2$; this is the adjoint representation.

One can find the weights of φ_λ by Klimyk's or by Freudenthal's formula. Klimyk's formula can be described "geometrically" as follows: For any weight μ we have to look at the weights $\mu + \delta - S\delta$ and the signs $\det S$. From the Cartan-Stiefel diagram for A_2 we copy the six vectors $\delta - S\delta$ on a small transparent (plastic) plate, with common origin of course, and attach the signs $\det S$ to them. We move the plate so that its origin coincides with the weight μ, and find the multiplicity m_μ as the signed sum of the multiplicities at the tips of the six vectors plus the value ϵ. (As regards the latter, one should begin the operation by determining the shifted orbit of λ, with the appropiate signs.)

Finally we consider splitting tensor products. Looking at weights works well in these simple cases. For instance:

The weights of $\varphi_{2\lambda_1}$ are $2\lambda_1 = 2\omega_1$ (the extreme weight), $2\omega_1-(\omega_1-\omega_2) = \omega_1+\omega_2 = -\omega_3, -\omega_3-(\omega_1-\omega_2) = 2\omega_2$ (these three weights form the $(\omega_1-\omega_2)$-string of $2\lambda_1$) and $-\omega_1, 2\omega_3, -\omega_2$ (e.g. by invariance under the Weyl group = all permutations of the ω_i). The tensor product $\Lambda_1 \otimes \Lambda_1$ has as weights all $\omega_i + \omega_j$ with $1 \leq i, j \leq 3$. The maximal weight is $2\omega_1 = 2\lambda_1$; thus $\varphi_{2\lambda_1}$ splits off, as the Cartan product of Λ_1 and Λ_1. The weights of $\Lambda_1 \otimes \Lambda_1$ are those of $\varphi_{2\lambda_1}$ and $-\omega_1, -\omega_2, -\omega_3$. The latter are the weights of Λ_2. Thus we have the splitting $\Lambda_1 \otimes \Lambda_1 = \varphi_{2\lambda_1} + \Lambda_2$.

Similarly $\Lambda_1 \otimes \Lambda_2$ has as weights all $\omega_i - \omega_j, i \neq j$, and 0 with multiplicity 3. These are the weights of $\varphi_{\lambda_1+\lambda_2} = \text{ad}$, with one weight 0 left over. This means $\Lambda_1 \otimes \Lambda_2 = \text{ad} + \varphi_0$.

For our A_2 a more explicit description of the irreps is of value (cf.[7]). We abbreviate $\mathbb{C}^3$ to V. Our Lie algebra being $\mathfrak{sl}(V)$, we can identify $V \wedge V$ with $V^\top$ equivariantly. Namely we identify $\bigwedge^3 V$ with $\mathbb{C}$ by sending $e_1 \wedge e_2 \wedge e_3$ to 1; then the $\wedge$-pairing of V and $V \wedge V$ to $\bigwedge^3 V$ becomes identified with the duality pairing of V and $V^\top$ to $\mathbb{C}$. The natural rep of A_2 in V is the fundamental rep to the fundamental weight λ_1. The induced rep in $V \wedge V$ is the one for λ_2; we see now that it equals the dual rep $\varphi_1^\triangle$ in $V^\top$. To describe the irrep φ_λ with $\lambda = n_1\lambda_1 + n_2\lambda_2$ we first form the tensor product $S^{n_1}V \otimes S^{n_2}V^\top$ (with the induced rep of course) (here S^m means the symmetric tensors in the m-fold tensor product), and then (assuming both n_1 and n_2 positive) take the trace (i.e., the map $V \otimes V^\top \to \mathbb{C}$ by $v \otimes \mu \to \mu(v)$) for any one of the V-factors and any one of the $V^\top$-factors. This sends the above space onto $S^{n_1-1}V \otimes S^{n_2-1}V^\top$. The induced rep on the kernel of the map is precisely φ_λ; note that (a) the highest weight occurring is $n_1\lambda_1 + n_2\lambda_2$, which does not occur in the image space), and that (b) it is easily verified that the dimension of the kernel agrees with that given for φ_λ by the Weyl dimension formula developed above. We denote this space and irrep also by $[n_1, n_2]$.

Let $\mu, = m_1\lambda_1 + m_2\lambda_2$, be a second weight. There is a fairly efficient algorithm for decomposing the tensor product $[n_1, n_2] \otimes [m_1, m_2]$ into its irreducible constituents. It is a two-stage process.

We introduce *intermediate* spaces (and reps) $[a_1, a_2, b_1, b_2]$, defined as the subspace of $S^{a_1}V \otimes S^{a_2}V^\top \otimes S^{b_1}V \otimes S^{b_2}V^\top$ on which all possible traces are 0. (A trace pairs some factor V with some factor $V^\top$ to $\mathbb{C}$, and sends the whole space to the tensor product of all the other factors.)

The first stage is a decomposition of $[n_1, n_2] \otimes [m_1, m_2]$ into intermediate spaces.

PROPOSITION A. *The rep $[n_1, n_2] \otimes [m_1, m_2]$ is equivalent to the direct sum of the $[n_1 - i, n_2 - j, m_1 - j, m_2 - i]$ for $0 \leq i \leq \min(n_1, m_2)$ and $0 \leq j \leq \min(n_2, m_1)$.*

The source for this is the distinguished element $\Sigma e_i \otimes \omega_i$ of $V \otimes V^\top$, where $\{e_i\}$ and $\{\omega_j\}$ are dual bases of V and $V^\top$. It does not depend on the choice of bases; it corresponds to id under the usual isomorphism between $V \otimes V^\top$ and the space $L(V, V)$ of linear maps from V to itself (or to the tensor δ_i^j in coordinate notation). (All this holds for any V, not just for $\mathbb{C}^3$.) We denote the element by TR and call it the *dual trace*. Now $V \otimes V^\top$ splits into the direct sum of the space of elements of trace 0 and the (one-dimensional) space spanned by TR, and this splitting is invariant under the action of $\mathfrak{gl}(V)$. This generalizes: In $V^n \otimes (V^\top)^m$ (where exponents mean tensor powers) we have the subspace W_0 of the tensors with all traces 0. Now in the product of the i-th V and j-th $V^\top$ we take the element TR and multiply it by arbitrary elements in the remaining factors. This produces a subspace, say U_{ij}, isomorphic to $V^{n-1} \otimes (V^\top)^{m-1}$. The sum (not direct!) of all the U_{ij} is a complement to W_0. Each U_{ij}, being of the same type as the original space, can be decomposed by the same process into the space of tensors with all traces 0 and a complement, generated by the TR's. The 0-trace tensors, for all i, j, give a subspace W_1. Continuing this way, one arrives at a decomposition of $V^n \otimes (V^\top)^m$ into a direct sum $W_0 \oplus W_1 \oplus \cdots$, where the terms in W_r are products of r TR's and a tensor with all traces 0, or sums of such. This decomposition is invariant under the action of $GL(V)$, and also under the symmetry group that consists in interchanging the V-factors and (independently) the $V^\top$-factors. (Cf.[25], p.150.) Applying this construction one proves Proposition A; we shall not go into the details.

The second stage consists in decomposing each intermediate space into irreps.

PROPOSITION B. *The space (and rep) $[n_1, n_2, m_1, m_2]$ is equivalent to the direct sum of the irreps $[n_1+m_1, n_2+m_2]$, $[n_1+m_1-2i, n_2+m_2+i]$ for $1 \le i \le \min(n_1, m_1)$, and $[n_1+m_1+j, n_2+m_2-2j]$ for $1 \le j \le \min(n_2, m_2)$.*

This depends very much on the fact that we are working in dimension 3 (i.e., $V = \mathbb{C}^3$). In this dimension we can identify $V \wedge V$ and $V^\top$ (and $\mathfrak{sl}(3, \mathbb{C})$-equivariantly so): Concretely, with dual bases $\{e_i\}$ and $\{\omega_j\}$ for V and $V^\top$ we send $e_1 \wedge e_2$ to ω_3 etc. (Abstractly, we can identify $V \wedge V \wedge V$ with $\mathbb{C}$, since it has dimension one and our Lie algebra acts trivially, and then the pairing of $V \wedge V$ and V to $V \wedge V \wedge V = \mathbb{C}$ shows that $V \wedge V$ acts as $V^\top$.) Let α be the map $V \otimes V \to V \wedge V \to V^\top$. The crucial fact is the following somewhat unexpected lemma.

LEMMA C. *Under $\alpha \otimes \mathrm{id}$ the subspace W_0 of $V \otimes V \otimes V^\top$ consisting of the tensors with both traces 0 maps to the symmetric elements in $V^\top \otimes V^\top$.*

In fact, more is true: an element, for which the two traces (to V) are equal, goes to a symmetric element. To show this we compose $\alpha \otimes \mathrm{id}$ with the map $\beta : V^\top \otimes V^\top \to V^\top \wedge V^\top \to V$ (the analog of α) and verify that this identical with the map "difference of the two traces". E.g., for $e_1 \otimes e_2 \otimes \omega_2$ the traces are 0 and e_1, and the other map has $e_1 \otimes e_2 \otimes \omega_2 \to e_1 \wedge e_2 \otimes \omega_2 \to \omega_3 \otimes \omega_2 \to \omega_3 \wedge \omega_2 \to -e_1$. $\checkmark$

Now to Proposition B. We consider two V's, not from the same symmetric product, and apply the map α to them. The kernel of this is clearly the space $[n_1 + m_1, n_2, 0, m_2]$ (since the kernel of α is the symmetric subspace S^2V, which combines the two symmetric products of V's into one long one). The image on the other hand is $[n_1 - 1, n_2 + m_2 + 1, m_1 - 1, 0]$, since by the lemma the two symmetric products of $V^\top$'s become one long one (in fact longer by one factor). (In the case $n_2 = m_2 = 0$ it is not quite obvious, but still true, that the traces are 0 here.) Iteration of this process yields Proposition B. Note $[a, b, 0, 0] = [0, 0, a, b] = [a, 0, 0, b] = [0, b, a, 0] = [a, b]$. $\surd$

As an example we take $[1,1] \otimes [1,1]$. ($[1,1]$ is the adjoint rep, of dimension 8.) By Proposition A we have

$$[1,1] \otimes [1,1] = [1,1,1,1] + [0,1,1,0] + [1,0,0,1] + [0,0,0,0].$$

By Proposition B we have

$$[1,1,1,1] = [2,2] + [0,3] + [3,0].$$

So finally

$$[1,1] \otimes [1,1] = [2,2] + [3,0] + [0,3] + [1,1] + [1,1] + [0,0].$$

$[0,0]$ is of course the trivial rep. One sees easily from the algorithm that $[n_1, n_2] \otimes [m_1, m_2]$ contains $[0,0]$ in its splitting iff $n_1 = m_2$ and $n_2 = m_1$. $\surd$

As an example for Brauer's algorithm (Prop.I of S3.7) we consider again A_2, i.e., $\mathfrak{sl}(2, \mathbb{C})$, with its two fundamental reps Λ_1 and Λ_2 (see §3.5) (where Λ_1 is the same as the $[1,0]$ above, i.e., the rep of $\mathfrak{sl}(2, \mathbb{C}$ by itself). We decompose $\Lambda_1 \otimes \Lambda_1$. The weights of Λ_1 are of course ω_1, ω_2, and ω_3; in terms of the fundamental weights λ_1 and λ_2 they are, resp, λ_1,$\lambda_2 - \lambda_1$, and $-\lambda_2$. Thus the character is $\chi_1 = e_{\lambda_1} + e_{\lambda_2 - \lambda_1} + e_{-\lambda_2}$. Brauer's algorithm asks us to form the product $\chi_1 \cdot A_{\lambda_1 + \delta}$, where δ is $\lambda_1 + \lambda_2$, and tells us that the result is $A_{2\lambda_1 + \delta} + A_{\lambda_2 + \delta} + A_{2\lambda_1}$. The third term is 0, because $2\lambda_1$ is singular (see §3.8, Prop.A). Dividing by A_δ and applying Weyl's character formula again, we find

$$\chi_1 \cdot \chi_1 = \chi_{2\lambda_1} + \chi_{\lambda_2} \text{ or } \varphi_1 \otimes \varphi_1 = \varphi_{2\lambda_1} \oplus \varphi_{\lambda_2},$$

agreeing with our earlier result above.— In terms of propositions A and B we also have $[1,0] \otimes [1,0] = [1,0,1,0] = [2,0] + [0,1]$.

3.10 The character ring

We return to the general semisimple $\mathfrak{g}$, and describe an important fact about the representation ring $R\mathfrak{g}$.

To each rep φ, or better to its equivalence class $[\varphi]$, is assigned its character χ_φ. By linearity this extends to a homomorphism of the free Abelian group F (see §3.5) into the additive group of the group ring $\mathcal{G}$ or $\mathbb{Z}\mathcal{I}$ (see §3.6). Since the character is additive on direct sums, the subgroup N (loc.cit.) goes to 0, and there is an induced additive homomorphism, say $\chi^\sim$, of $R\mathfrak{g}$ into $\mathcal{G}$. The character is multiplicative on tensor products, and so $\chi^\sim$ is in fact a ring homomorphism. The characters χ_λ of the irreps φ_λ associated to the dominant weights λ are linearly independent in $\mathcal{G}$, e.g. by Weyl's formula: The skew elemens $A_{\lambda+\delta} = \chi_\lambda \cdot A_\delta$ are independent. In fact, since the $A_{\lambda+\delta}$ are a basis for the additive group of skew elements in $\mathcal{G}$, the χ_λ are an additive basis for the ring of symmetric elements. We state this as

PROPOSITION A. *The map $\chi^\sim$ is an isomorphism of $R\mathfrak{g}$ onto the subring $\mathcal{G}^\mathcal{W}$ of $\mathcal{G}$ formed by the symmetric elements, the invariants of the Weyl group.*

$\mathcal{G}^\mathcal{W}$ (and then also $R\mathfrak{g}$) is called the *character ring* of $\mathfrak{g}$. (We recall that $R\mathfrak{g}$ is a polynomial ring; see §3.5, Proposition A.)

We consider a slight generalization of $R\mathfrak{g}$:

The character $\chi = \chi^\sim(\varphi)$ of a rep φ can be considered as a function on $\mathfrak{h}_0$. At each t in the co-root lattice $\mathcal{T}$ it takes the value d_φ = degree of φ, since then $\exp(2\pi i\lambda(t)) = 1$ for all weights λ. Now it may happen for some φ that there are other t in $\mathfrak{h}_0$ where the character takes this value d_φ (i.e., where all the weights in χ take integral values). All these points clearly form a lattice $\mathcal{L}$ (depending on φ) in $\mathfrak{h}_0$; here we assume φ faithful, i.e., no simple constituent of $\mathfrak{g}$ goes to 0 under φ. The lattice is of course invariant under the Weyl group. Furthermore it contains $\mathcal{T}$ (of course), and is contained in the center lattice $\mathcal{Z}$ because every root appears as the difference of two weights of φ, as one can see from Theorem B(c) of §3.2 (note that by Proposition D, §2.6 no root α can be orthogonal to all the weights of φ). [The significance of all this is the following: With $\mathfrak{g}$ is associated a simply connected compact Lie group G, whose Lie algebra is the compact form $\mathfrak{u}$ of $\mathfrak{g}$. The torus $i\mathfrak{h}_0/2\pi i\mathcal{T}$ becomes identified with a subgroup (the torus $\mathbb{T}$) of G; the finite group $2\pi i\mathcal{Z}/2\pi i\mathcal{T}$ becomes the center of G. The rep φ of $\mathfrak{g}$ generates a rep φ^G of G, which to the element represented by $2\pi iH$ assigns the operator $\exp(2\pi i\varphi(H))$. The elements with H in $\mathcal{T}$, which correspond to 1 in G, go to id under φ^G. The kernel N of $\varphi^\sim$ is a subgroup of the center of G; its inverse image in $i\mathfrak{h}_0$ is precisely the lattice $2\pi i\mathcal{L}$. Thus a rep φ whose character χ_λ takes the value d_λ on $\mathcal{L}$ corresponds to a rep of G that factors through the quotient G/N.]

To a given lattice $\mathcal{L}$ between $\mathcal{T}$ and $\mathcal{Z}$ and invariant under $\mathcal{W}$ we associate all the reps φ whose character takes the value d_φ on $\mathcal{L}$. A direct sum of reps is of this type iff all the summands are. We now construct the representation ring $R\mathfrak{g}_{\mathcal{L}}$ for this set of reps by the same recipe by which we constructed the ring $R\mathfrak{g}$. It is clear that this is a subring of $R\mathfrak{g}$, spanned additively by the irreps with the property at hand. It may fail to be a polynomial ring as we will see by an example below.

Example 1: B_l.

Here we take as $\mathcal{L}$ the only possibility (outside of $\mathcal{T}$), namely $\mathcal{Z}$ itself (cf. §2.5). [This amounts to considering reps of the orthogonal group $SO(2l+1)$ rather then reps of the corresponding simply connected group, the so-called *spin group* $Spin(2l+1)$.] The crucial element at which we have to evaluate the characters is the vector e_1. In order for all the exponentials in χ_λ to have value 1 at e_1, the coefficient f_1 of $\lambda = \Sigma f_i \omega_i$ must be integral (and not half-integral). Writing $\lambda = \Sigma n_i \lambda_i$, in terms of fundamental reps, this means that n_l must be even, i.e., λ must be a non-negative-integral linear combination of $\lambda_1, \ldots, \lambda_{l-1}$ and $2\lambda_l$, and that $R\mathfrak{g}_{\mathcal{L}}$ is the subring of $R\mathfrak{g}$ generated by $\Lambda_1, \Lambda_2, \ldots, \Lambda_{l-1}$ and $\varphi_{2\lambda_l}$. [$\varphi_{2\lambda_l}$ is in fact $\Lambda^l \mathfrak{o}(2l+1, \mathbb{C})$; we also write Λ_l for it.]

Now all the exponentials in the character of the spin rep $\Delta = \varphi_l$ take value -1 at e_1. Therefore $\Delta \otimes \Delta$ is in our subset, and we have an equation $\Delta \otimes \Delta = \varphi_{2\lambda_l} + \cdots$, where the dots represent a sum of terms $\Lambda_1, \ldots, \Lambda_{l-1}$. (Details below.) This means that $R\mathfrak{g}_{\mathcal{L}}$ is the subring of $R\mathfrak{g}$ generated by $\Lambda_1, \Lambda_2, \ldots, \Lambda_{l-1}$ and $\Delta \otimes \Delta$, and therefore it is a polynomial ring (with these elements - or with $\Lambda_1, \Lambda_2, \ldots, \Lambda_l$ - as generators).

Example 2: D_l.

Here there are several possibilities for $\mathcal{L}$. We choose the lattice generated by $\mathcal{T}$ and the vector e_1 [this again corresponds to reps of $SO(2l)$ rather then of the simply connected group $Spin(2l)$].

Again the λ in question must have the form $\Sigma f_i \omega_i$ with f_1 integral or, in the form $\Sigma n_i \lambda_i$, with $n_{l-1} + n_l$ even. Introducing $\lambda' = \lambda_{l-1} + \lambda_l = \omega_1 + \omega_2 + \cdots + \omega_{l-1}, \lambda^+ = 2\lambda_l = \omega_1 + \omega_2 + \cdots + \omega_l$, and $\lambda^- = 2\lambda_{l-1} = \omega_1 + \omega_2 + \cdots + \omega_{l-1} - \omega_l$ (and denoting the corresponding reps by Λ', Λ^+ and Λ^-), we can describe these λ as the non-negative-integral linear combinations of $\lambda_1, \ldots, \lambda_{l-2}, \lambda', \lambda^+$, and λ^-. The ring $R\mathfrak{g}_{\mathcal{L}}$ is then the subring of $R\mathfrak{o}(2l)$ generated by $\Lambda_1, \ldots, \Lambda_{l-1}, \Lambda^+, \Lambda^-$.

Now the tensor products of the spin reps Δ^+ and Δ^- split according to $\Delta^+ \otimes \Delta^+ = \Lambda^+ + \cdots, \Delta^- \otimes \Delta^- = \Lambda^- + \cdots, \Delta^+ \otimes \Delta^- = \Lambda_{l-1} + \cdots$, where the dots in all three cases represent a sum of terms $\Lambda_1, \ldots, \Lambda_{l-2}$, as one can see from the weights (details below). This means that the subring $R\mathfrak{g}_{\mathcal{L}}$ of $R\mathfrak{g}$ is also generated by $\Lambda_1, \ldots, \Lambda_{l-2}, \Delta^+ \otimes \Delta^+, \Delta^- \otimes \Delta^-$ and $\Delta^+ \otimes \Delta^-$. This is not a polynomial ring; as regards Δ^+ and Δ^- it is of the form $A[x^2, y^2, xy]$, which in turn can be described as $A[u, v, w]/(uv - w^2)$.

Details and comments.

For B_l the exact equation is $\Delta \otimes \Delta = \varphi_{2\lambda_l} + \Lambda_{l-1} + \Lambda_{l-2} + \cdots + \Lambda_0$, as one can see by properly distributing the weights of $\Delta \otimes \Delta$ (here Λ_0 is, as always, the trivial rep).

We indicate the argument: The highest weight of $\Delta \otimes \Delta$ is $\omega_1 + \omega_2 + \cdots + \omega_l$; thus Λ_l occurs, once, as the Cartan product. The weight $\omega_1 + \omega_2 + \cdots + \omega_{l-1}$ occurs twice in $\Delta \otimes \Delta$, but only once in Λ_l; it is the highest of the weights of $\Delta \otimes \Delta$ after those of Λ_l have been removed. Thus Λ_{l-1} occurs, once. Next we look at $\omega_1 + \omega_2 + \cdots + \omega_{l-2}$. It occurs four times in $\Delta \otimes \Delta$, twice in Λ_l and once in Λ_{l-1}; this forces Λ_{l-2} to be present, once. Etc.

For D_l the situation is of course a bit more complicated. Λ', as defined above, is $\bigwedge^{l-1} \mathfrak{o}(2l, \mathbb{C})$, since $\omega_1 + \omega_2 + \cdots + \omega_{l-1}$ is the highest weight of the latter, and Λ_{l-1} has the right dimension, from the degree formula. But $\bigwedge^l \mathfrak{o}(2l, \mathbb{C})$ ($= \Lambda_l$ in short) is not irreducible; it splits in fact into the direct sum of the Λ^+ and Λ^- introduced above. This comes about through the so-called (Hodge) *star*-operation $*$: For a complex vector space V of dimension n, with a non-degenerate quadratic form $(\cdot, \cdot)$ and a given "volume element" u (element of $\bigwedge^n V$ with $(u, u) = 1$) this is the map from $\bigwedge^p V$ to $\bigwedge^{n-p} V$ defined by $v \wedge *w = (v, w)u$. One verifies that $*$ is equivariant wr to the operators induced in $\bigwedge^p V$ and $\bigwedge^{n-p} V$ by the elements of the orthogonal Lie algebra associated with $(\cdot, \cdot)$.

For $\mathbb{C}^n$ with metric Σx_i^2 and $u = e_1 \wedge e_2 \wedge \ldots \wedge e_n$ this sends $e_{i_1} \wedge e_{i_2} \wedge \ldots \wedge e_{i_p}$ to $e_{j_1} \wedge e_{j_2} \wedge \ldots \wedge e_{j_{n-p}}$, where the j's form the complement to the i's in $\{1, 2, \ldots, n\}$ and are so ordered that $\{i_1, i_2, \ldots, i_p, j_1, j_2, \ldots, j_{n-p}\}$ is an even permutation of $\{1, 2, \ldots, n\}$. (Note that in our description of B_l and D_l in §2.14 we use a different metric form.)

Clearly the operator $**$ on $\Lambda^p V$ is the scalar map $(-1)^{p(n-p)} id$. In particular, for our case D_l with $n = 2l$ and taking $p = l$, the $*$-map sends $\bigwedge^l \mathbb{C}^{2l}$ to itself, and its square is $(-1)^l$. Thus the eigenvalues of $*$ on this space are ± 1 for l even and $\pm i$ for l odd, and the space splits into the corresponding eigenspaces. [An improper orthogonal (wr to $(\cdot, \cdot)$) transformation. e.g. $\operatorname{diag}(1, \ldots, 1, -1)$, interchanges the two eigenspaces, which therefore have the same dimension.] As noted above for the general case, $*$ is equivariant wr to the action of Λ_l on $\bigwedge^l \mathbb{C}^{2l}$. Therefore the eigenspaces of $*$ go into themselves under Λ_l, and this is the promised splitting of Λ_l into Λ^+ and Λ^-. The exact equations are now $\Delta^+ \otimes \Delta^+ = \Lambda^+ + \Lambda_{l-2} + \Lambda_{l-4} + \ldots$, $\Delta^- \otimes \Delta^- = \Lambda^- + \Lambda_{l-2} + \Lambda_{l-4} + \ldots$, $\Delta^+ \otimes \Delta^- = \Lambda_{l-1} + \Lambda_{l-3} + \Lambda_{l-5} + \ldots$, each sum ending in Λ_1 or Λ_0, as one can see again by enumerating the weights.

3.11 Orthogonal and symplectic representations

The purpose of this section is to decide which representations of the various semisimple Lie algebras consist, in a suitable coordinate system, of orthogonal matrices, resp of symplectic matrices. The results are due to I.A. Mal'cev [20]. We follow the argument by A.K. Bose and J. Patera [1].

First some linear algebra.

Let V be a vector space (over $\mathbb{F}$, of finite dimension). We write $B(V)$ for the vector space of bilinear functions from $V \times V$ to $\mathbb{F}$, and $L_*(V)$ for the vector space of all linear maps from V to its dual space $V^\top$. There is a canonical isomorphism between $B(V)$ and $L_*(V)$: Let b be a bilinear form; the corresponding map $b' : V \to V^\top$ sends a vector v into that linear function on V whose value at any w is $b(v, w)$. In other words, we get b' from b by fixing the first variable. The dual of b' is also a map from V to $V^\top$ (in reality it is a map from $V^{\top\top}$ to $V^\top$; but $V^{\top\top}$ is canonically identified with V). Of course this dual is nothing but the map obtained from b by fixing the second variable; i.e., we have $b''(v)(w) = b(w, v)$. Thus b is symmetric, resp. skew-symmetric, if b' equals its dual, resp. equals the negative of its dual. Also, b is non-degenerate exactly when b' (or b'') is invertible.

Let A be an operator on V. We let A operate on $V^\top$ as $A^\triangle = -A^\top$; that is, we define $A^\triangle \rho(v) = -\rho(Av)$ for any ρ in $V^\top$ and v in V. We use this *infinitesimal contragredient* with the applications to contragredient representations of Lie algebras in mind. We also let A operate on $B(V)$ by $Ab(v, w) = -b(Av, w) - b(v, Aw)$. The isomorphism of $B(V)$ and $L_*(V)$ then makes Ab correspond to the map $A^\triangle \circ b' - b' \circ A$ from V to $V^\top$. In particular, b is (infinitesimally) invariant under A (i.e., $b(Av, w) + b(v, Aw)$ is identically 0) iff b' is an A-equivariant map from V to $V^\top$ (i.e., satisfies $A^\triangle \circ b' = b' \circ A$).

Let now $\mathfrak{g}$ be a Lie algebra and let φ be a representation of $\mathfrak{g}$, on the vector space V. Associated to φ are then the representations on $V^\top$, on $B(V)$ and on $L_*(V)$, obtained by applying to each operator $\varphi(X)$ the constructions of the preceding paragraph. The representation on $V^\top$ is the contragredient or dual to φ, denoted by $\varphi^\triangle$. We will be particularly interested in the φ-invariant bilinear forms, i.e. the elements b of $B(V)$ that satisfy $b(\varphi(X)v, w) + b(v, \varphi(X)w) = 0$ for all v, w in V and X in $\mathfrak{g}$. From the discussion above we see that under the isomorphism of $B(V)$ with $L_*(V)$ they correspond to the φ-equivariant maps from V to $V^\top$, i.e. the linear maps $f : V \to V^\top$ with $f \circ \varphi(X) = \varphi(X)^\triangle \circ f$ for all X in $\mathfrak{g}$.

We come to our basic definitions: The representation φ of the Lie algebra $\mathfrak{g}$ on the space V is called *self-contragredient* or *self-dual* if it is equivalent to its contragredient $\varphi^\triangle$. This amounts to the existence of a φ-equivariant isomorphism from V to $V^\top$, or, in view of our discussion above, the existence of a non-degenerate invariant (i.e. infinitesimally invariant under all $\varphi(X)$) bilinear form on V. One calls φ *orthogonal* if there exists a non-degenerate symmetric bilinear form on V, invariant (infinitesimally of course) under all $\varphi(X)$. Similarly φ is called *symplectic* if there exists a non-degenerate skew bilinear form on V, invariant under all $\varphi(X)$. Another way to say this is that all $\varphi(X)$ belong to the orthogonal Lie algebra defined by the symmetric form, resp. to the symplectic Lie algebra defined by the skew form.

No uniqueness is required in this definition; there might be several linearly independent invariant forms; φ could even be orthogonal and symplectic at the same time. The situation is different however, if the underlying field is $\mathbb{C}$ (as we shall assume from now on) and φ is irreducible.

PROPOSITION A. *Let φ be over $\mathbb{C}$ and irreducible. Then*

(a) a φ-invariant bilinear form is either non-degenerate or 0;

(b) up to a constant factor there is at most one non-zero invariant bilinear form;

(c) a φ-invariant bilinear form is automatically either symmetric or skew (but not both).

To restate this in a slightly different form, we note first that the space $B(V)$ is isomorphic (and φ-equivariantly so) to the tensor product $V^\top \otimes V^\top$. (An element $\lambda \otimes \mu$ of the latter defines a bilinear form by $\lambda \otimes \mu(v,w) = \lambda(v) \cdot \mu(w)$.) Under this correspondence symmetric (resp skew) forms correspond to symmetric (resp skew) elements of $V^\top \otimes V^\top$.

PROPOSITION A′. *Let φ be over $\mathbb{C}$ and irreducible.*

(a,b) The space of invariants of $\varphi^\triangle \otimes \varphi^\triangle$ in $V^\top \otimes V^\top$ is of dimension 0 *or* 1, *the latter exactly if φ is self-dual;*

(c) A (non-zero) self-dual φ is either orthogonal or symplectic (but not both); it is orthogonal if the second symmetric power $S^2\varphi$ has an invariant (i.e., contains the trivial representation), and is symplectic if the second exterior power $\bigwedge^2 \varphi$ has an invariant.

Proof (of A and A′): we look at an invariant bilinear form as an equivariant map from V to $V^\top$. Since $\varphi^\triangle$, on $V^\top$, is of course also irreducible, Schur's lemma gives the result (a). For (b): If b_1 and b_2 are two invariant bilinear forms, then for a suitable number k the form $b_1 - kb_2$ is degenerate (we are over $\mathbb{C}$) and still invariant; now apply (a). For (c): A bilinear form b is, uniquely, the sum of a symmetric and a skew one [by $b(v,w) = 1/2(b(v,w) + b(w,v)) + 1/2(b(v,w) - b(w,v))$]. (In other words, we have the invariant decomposition $V^\top \otimes V^\top = S^2V^\top + \bigwedge^2 V^\top$.) If b is invariant, so are its symmetric and its skew parts; now apply (b). $\surd$

We need a few more obvious general formal facts.

PROPOSITION B.

(a) If φ is orthogonal [resp symplectic], then the dual $\varphi^{\triangle}$ is also orthogonal [resp symplectic]; the r-th exterior power $\bigwedge^r \varphi$ is orthogonal [resp symplectic] for odd r and orthogonal for even r;

(b) the direct sum of two orthogonal [resp symplectic] representation is orthogonal [resp symplectic] ;

(c) the tensor product of two orthogonal or two symplectic representations is orthogonal;

(d) the tensor product of an orthogonal and a symplectic representation is symplectic.

The proof of this, using equivariant maps from V to $V^{\top}$ and natural identifications such as dual of exterior power = exterior power of the dual, is straightforward. E.g. for (a): If φ is orthogonal, there is a φ-equivariant isomorphism from V to $V^{\top}$ equal to its dual; the inverse of this map is then an equivariant map from $V^{\top}$ to $V^{\top\top} = V$, equal to its dual. √

For the reducible case we need a simpleminded lemma.

LEMMA C. *Suppose the rep φ of $\mathfrak{g}$ on V is direct sum of irreducible reps φ_i on subspaces V_i, and φ_1, on V_1, is not the dual of any of the other φ_i, $i > 1$. Then, if φ is orthogonal [resp symplectic], so is φ_1.*

Proof: First, $\varphi^{\triangle}$ on $V^{\top}$ is of course direct sum of the $\varphi_i^{\triangle}$ on the $V_i^{\top}$. An equivariant isomorphism b' from V to $V^{\top}$ gives then a similar map b'_1 from V_1 to $V_1^{\top}$, by the hypothesis on φ_1, making φ_1 self-dual. If the dual of b' is $\pm b'$, the same holds for b'_1. √

We return now to our semisimple Lie algebra $\mathfrak{g}$, with all its machinery (§2.2 ff.). We are given a dominant weight λ and the associated irrep φ_λ with λ as extreme weight. There is a simple criterion for self-duality in terms of weights.

PROPOSITION D. *φ_λ is self-dual iff its minimal weight is $-\lambda$.*

Proof: "Minimal" is of course understood rel the order in $\mathfrak{h}_0^{\top}$ that we have been using all along. — The definition of the contragredient, $\varphi^{\triangle}(X) = -\varphi(X)^{\top}$, shows that the weights of $\varphi_\lambda^{\triangle}$ are the negatives of those of φ_λ. Thus $\lambda^{\triangle}$, the extreme and maximal weight of $\varphi_\lambda^{\triangle}$, is the negative of the minimal weight of φ_λ. (Changing the sign reverses the order in $\mathfrak{h}_0^{\top}$.) √

There are other ways to look at this. By considering the reversed order in $\mathfrak{h}_0^{\top}$ one sees easily that the minimal weight of φ_λ is that element of the Weyl group orbit of λ that lies in the negative of the closed fundamental

Weyl chamber. It is therefore the image of λ under the opposition element *op* of $\mathcal{W}$ (see §2.15). Thus Proposition D can be restated as

PROPOSITION D′. *φ_λ is self-dual iff the opposition sends λ to $-\lambda$.*

This is of course automatic if the opposition is $-$id; in other words, if $\mathcal{W}$ contains the element $-$id.

We come now to our main task, the discussion of the individual simple Lie algebras. In each case we shall indicate for each dominant weight λ whether φ_λ is self-dual, and if so, whether it is orthogonal or symplectic. We write the λ's as $\Sigma f_i\omega_i$ (as in §3.5), and also as $\Sigma n_i\lambda_i$ (in terms of the fundamental weights λ_i and non-negative integral n_i). One often describes such a λ by attaching the integer n_i to the vertex α_i in the Dynkin diagram. The result is contained in the following long theorem. (The trivial representation is of course self-dual and orthogonal.)

THEOREM E.

(a) A_l : φ_λ is self-dual iff $f_1 = f_2 + f_l = f_3 + f_{l-1} = \cdots$ (equivalently $n_1 = n_l, n_2 = n_{l-1}, \ldots$) [and thus for all λ in the case $l = 1$]. It is then symplectic if $l \equiv 1 \bmod 4$ and f_1 odd ($n_{(l+1)/2}$, the middle n_i, odd), and orthogonal otherwise;

(b) B_l : φ_λ is always self-dual. It is symplectic if $l \equiv 1$ or $2 \bmod 4$ and the f_i are half-integral (n_l is odd), and orthogonal otherwise;

(c) C_l : φ_λ is always self-dual. It is symplectic if Σf_i is odd ($n_1 + n_3 + n_4 + \cdots$ is odd), and orthogonal otherwise;

(d) Dl : φ_λ is self-dual iff either l is even or l is odd and $f_l = 0 (n_{l-1} = n_l)$. It is then symplectic if $l \equiv 2 \bmod 4$ and the f_i are half-integral ($n_l - 1 + n_l$ is odd), and orthogonal otherwise;

(e) G_2 : φ_λ is self-dual and orthogonal for every λ;

(f) F_4 : φ_λ is self-dual and orthogonal for every λ;

(g) E_6 : φ_λ is self-dual iff $\Sigma f_i/3 = f_1 + f_6 = f_2 + f_4 = f_3 + f_4$ ($n_1 = n_4$ and $n_2 = n_4$). It is then orthogonal (and the f_i are integers);

(h) E_7 : φ_λ is always self-dual. It is orthogonal if the f_i are integral ($n_1 + n_3 + n_7$ is even), and symplectic otherwise;

(i) E_8 : φ_λ is always self-dual and orthogonal .

We start with the question of self-duality, using Proposition D or D′. As we know from §2.15, the opposition is $-$id for the simple Lie algebras A_1, B_l, C_l , the D_l for even l , G_2, F_4, E_7, E_8; thus all their irreps are self-dual. There is a problem only for A_l with $l > 1$, for D_l with $l = 2k+1$ odd, and for E_6. For A_l the opposition is given by $\omega_i \to \omega_{l+2-i}$ (see loc.cit.).

Thus *op* sends the fundamental weight $\lambda_i = \omega_1 + \cdots + \omega_i$ to $\omega_{l+1} + \cdots + \omega_{l+2-i} = -\omega_1 - \cdots - \omega_{l+1-i} = -\lambda_{l+1-i}$ (we used $\Sigma_1^{l+1}\omega_i = 0$ here). The weight $\lambda = \Sigma n_i\lambda_i$ then goes to $-\lambda$ under *op* iff the relations $n_i = n_{l+1-i}$ hold. $\surd$

For D_{2k+1} the opposition sends ω_i to $-\omega_i$ for $1 \leq i \leq l-1$, and keeps ω_l fixed (see loc.cit.). It sends the fundamental weights λ_i to $-\lambda_i$ for $1 \leq i \leq l-2$, and sends λ_{l-1} to $-\lambda_l$ and λ_l to $-\lambda_{l-1}$. Thus a dominant weight $\lambda = \Sigma n_i\lambda_i$ gives a self-dual irrep iff $n_{l-1} = n_l$. $\surd$

For E_6 the opposition interchanges λ_1 and λ_5, and also λ_2 and λ_4 (see loc.cit.) Thus a dominant $\lambda = \Sigma n_i\lambda_i$ is self-dual iff $n_1 = n_5$ and $n_2 = n_4$. $\surd$

Now comes the question orthogonal vs. symplectic. We settle this first for A_1, whose representations we know from §1.11. Here the representation $D_{1/2}$ is symplectic (the determinant $x_1y_2 - x_2y_1$ of two vectors x, y in $\mathbb{C}^2$ is the relevant invariant skew form; or one notes that for a 2×2 matrix M the condition $M^\top J + JM = 0$ is identical with $\operatorname{tr} M = 0$). It follows now easily from Proposition B, Lemma C and the Clebsch-Gordan series (§1.12) that D_s is orthogonal for integral s (i.e., odd dimension) and symplectic for half-integral s (i.e., even dimension) (and also for $s = 0$).

The other simple Lie algebras will be handled with the help of a general result, which involves the notion of *principal three-dimensional sub Lie algebra* (abbreviated to *PTD*). Let $\mathfrak{g}$ be a semisimple Lie algebra, as before; we use the concepts listed at the beginning of this chapter. Since the fundamental roots $\alpha_1, \ldots, \alpha_l$ are a basis of $h_0^\top$, there exists a unique element H_p in $\mathfrak{h}_0$ (in fact in the fundamental Weyl chamber) with $\alpha_i(H_p) = 2$ for $1 \leq i \leq l$. We write H_p as Σp_iH_i, choose constants c_i, c_{-i} so that $c_i \cdot c_{-i} = p_i$, and introduce $X_p = \Sigma c_iH_i$ and $X_{-p} = \Sigma c_iH_i$. Using the relations $[HX_i] = \alpha_i(H)X_{-i}$, $[X_iX_{-i}] = H_i$, $[X_iX_{-j}] = 0$ for $i \neq j$ one verifies $[H_pX_p] = 2X_p$, $[H_pX_{-p}] = -2X_{-p}$, and $[X_pX_{-p}] = H_p$. The sub Lie algebra $\mathfrak{g}_p$ of $\mathfrak{g}$ spanned by H_p, X_p, X_{-p}, visibly of type A_1, is by definition a *PTD* of $\mathfrak{g}$. It has $\mathbb{C}H_p$ as Cartan sub Lie algebra (with the obvious order, which agrees with the order in $\mathfrak{h}_0 : H_p$ is > 0); its root system consists of $\pm\alpha_p$, defined by $\alpha_p(H_p) = 2$.

A representation φ of $\mathfrak{g}$ restricts to a representation $\varphi^\sim$ of $\mathfrak{g}_p$; since H_p lies in $\mathfrak{h}$, any weight ρ of φ restricts to a weight $\rho^\sim$ of $\varphi^\sim$, and all weights of $\varphi^\sim$ arise in this way. (Such a $\rho^\sim$ amounts of course simply to the eigenvalue $\rho(H_p)$ of H_p.)

In general $\varphi^\sim$ will not be irreducible, and will split into a sum of the irreps of $\mathfrak{g}_p$, i.e., into D_s's. We come to the property of $\varphi^\sim$ that we utilize for our problem.

LEMMA F. *Let $\varphi = \varphi_\lambda$ be an irrep of $\mathfrak{g}$, with extreme weight λ. Then*

(a) $\lambda^\sim$ is the maximal weight of $\varphi^\sim$ and has multiplicity 1;

(b) in the splitting of $\varphi^\sim$ the representation D_s with $2s = \lambda(H_p)$ (the top constituent) occurs exactly once.

Proof: The weights of φ, other than λ itself, are of the form $\rho = \lambda - \Sigma k_i \alpha_i$, with non-negative integers k_i and $\Sigma k_i > 0$. Thus from $\alpha_i(H_p) = 2$ we have $\rho(H_p) = \lambda(H_p) - 2\Sigma k_i < \lambda(H_p)$. Since λ has multiplicity 1 in φ, part (a) follows. Part (b) is an immediate consequence, since in any D_s the largest eigenvalue of H is precisely $2s$. $\surd$

We can now state our criterion.

PROPOSITION G. *φ_λ (assumed self-dual) is orthogonal if $\lambda(H_p)$ is even, and symplectic if $\lambda(H_p)$ is odd.*

Proof: Clearly $\varphi_\lambda^\sim$ is orthogonal if φ_λ is so, and symplectic if φ_λ is so. We apply Lemma C to $\varphi_\lambda^\sim$ and its splitting into D_s's. Since the top constituent of $\varphi_\lambda^\sim$ occurs only once, by lemma F, it follows from Lemma C that the top constituent is orthogonal if $\varphi_\lambda^\sim$ is so, and symplectic if $\varphi_\lambda^\sim$ is so. As we saw above in our discussion of the behavior of the D_s's, the top constituent of $\varphi_\lambda^\sim$ is orthogonal if $\lambda(H_p)$ is even, and symplectic if it is odd. $\surd$

(Incidentally, all eigenvalues of H_p under φ_λ are of the same parity, since all $\alpha(H_p)$ are even; thus any weight of φ_λ could be used in Proposition G.)

For the proof of Theorem E it remains to work this out for the simple Lie algebras.

With $H_p = \Sigma p_i H_i$ and $\lambda = \Sigma n_i \lambda_i$ the crucial value $\lambda(H_p)$ becomes $\Sigma n_i p_i$. The constants p_i are determined from $\alpha_i(H_p) = 2$, i.e. from $\Sigma a_{ij} p_j = 2$, where $a_{ij} = \alpha_i(H_j)$ are the Cartan integers. For the simple Lie algebras the a_{ij} are easily found from the α_i in §2.14 and the H_i in §3.5.

As an example, for G_2 we have $\alpha_1 = \omega_2, \alpha_2 = \omega_1 - \omega_2, H_1 = (-1, 2, -1)$, $H_2 = (1, -1, 0)$. Thus $a_{11} = 2, a_{12} = -1, a_{21} = -3, a_{22} = 2$. We get the equations $2p_1 - p_2 = 2, -3p_1 + 2p_2 = 2$, giving $p_1 = 6, p_2 = 10$. Thus H_p is $6H_1 + 10H_2$, and for $\lambda = n_1\lambda_1 + n_2\lambda_2$ we have $\lambda(H_p) = 6n_1 + 10n_2$. This is always even, in agreement with Theorem E, (e).

[We also note that the definition of H_p is dual to that for the lowest weight δ (see §3.1), except for a factor 2. Thus H_p can also be found as $\Sigma_{\alpha>0} H_\alpha$.]

We list the results in the usual way, attaching the coefficient p_i to the vertex α_i of the Dynkin diagram.

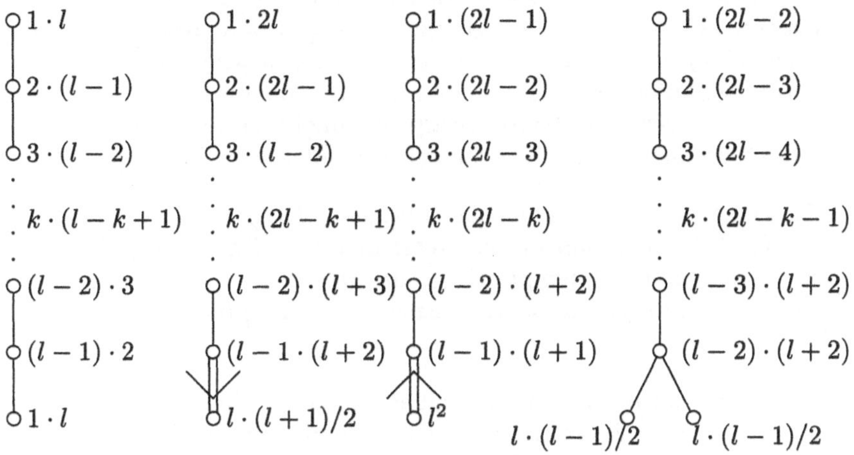

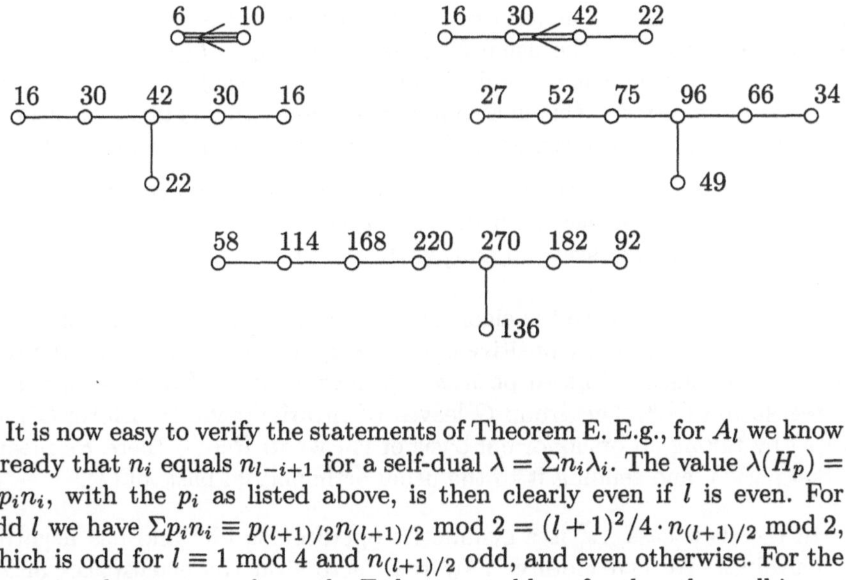

It is now easy to verify the statements of Theorem E. E.g., for A_l we know already that n_i equals n_{l-i+1} for a self-dual $\lambda = \Sigma n_i \lambda_i$. The value $\lambda(H_p) = \Sigma p_i n_i$, with the p_i as listed above, is then clearly even if l is even. For odd l we have $\Sigma p_i n_i \equiv p_{(l+1)/2} n_{(l+1)/2} \bmod 2 = (l+1)^2/4 \cdot n_{(l+1)/2} \bmod 2$, which is odd for $l \equiv 1 \bmod 4$ and $n_{(l+1)/2}$ odd, and even otherwise. For the exceptional cases note that only E_7 has any odd p_i; for the others all irreps are orthogonal.

As a minor application: The seven-dimensional rep of G_2 of §3.5 can be interpreted, using Theorem E(e), as giving an inclusion $G_2 \subset B_3$.This is the inclusion described in §2.14.

One reads off from Theorem E that for the following simple Lie algebras and only for these all representations are orthogonal:

$$\mathfrak{o}(n, \mathbb{C}) \text{ with } n \equiv \pm 1 \text{ or } 0 \bmod 8, G_2, F_4, E_8.$$

To conclude we discuss briefly the situation for compact Lie groups.

Let then G be a compact Lie group, and let φ be a representation of G on the (complex) vector space V. How does one decide whether φ is equivalent to a representation in $O(n)$ [i.e., by real orthogonal matrices], or in $Sp(n)$ [i.e., by unitary-symplectic matrices]? We might as well replace V by $\mathbb{C}^n$ and G by its image $\varphi(G)$. The answer is as follows.

THEOREM H. *A compact subgroup of $GL(n, \mathbb{C})$ is conjugate in $GL(n, \mathbb{C})$ to a subgroup of the (real) orthogonal group $O(n)$ [resp, for even n, the (unitary) symplectic group $Sp(n/2)$] , iff it leaves invariant a non-degenerate symmetric [resp skew-symmetric] bilinear form on $\mathbb{C}^n$.*

(The invariance is now meant in the global sense, $b(gv, gw) = b(v, w)$ for every g in G; not in the infinitesimal as earlier for Lie algebras. As before, symmetric or skew invariant forms correspond to invariants in the second symmetric or exterior power of V.)

Proof: We begin by finding a positive definite Hermitean form $\langle\cdot,\cdot\rangle$ on $\mathbb{C}^n$ that is invariant under G. The existence of this is a standard fact. A short proof (L. Auerbach) is as follows: Let G operate on the vector space of all Hermitean forms (by the usual formula $g \cdot h(v, w) = h(g^{-1} \cdot v, g^{-1} \cdot w)$). Let C stand for the convex hull of the set of G-transforms of some chosen positive definite form; this is a compact set, consisting entirely of positive definite forms, and is invariant under G. Its barycenter is the required $\langle\cdot,\cdot\rangle$.

Now let b be a symmetric [resp skew] form as of Theorem H. The equation $b(v, w) = \langle Av, w\rangle$ defines, as usual, a conjugate-linear automorphism A of $\mathbb{C}^n$, self-adjoint [resp skew-adjoint] wr to the positive definite form $Re\langle\cdot,\cdot\rangle$ on $\mathbb{C}^n_{\mathbb{R}} = \mathbb{R}^2n$.

In the symmetric case the eigenvalues of A are all real; because of $Aiv = -iAv$ there are as many positive ones as negative ones. Let W be the real span of the eigenvectors to positive eigenvalues; it is of dimension n and a real form of $\mathbb{C}^n$. The group G leaves W invariant. We transform G into $O(n)$ by taking an orthonormal basis of W, wr to $Re\langle\cdot,\cdot\rangle$ (here Re means "real part"), and sending it to the usual orthonormal basis of $\mathbb{R}^n$.

In the skew case, A^2 is a symmetric operator on $\mathbb{R}^{2n}$ and has negative eigenvalues. We can modify A by real factors on the eigenspaces of A^2 so that A^2 is $-\mathrm{id}$. For any unit vector v we have then $\langle Av, v\rangle = 0$ and $b(Av, v) = -1$. The space $((v, Av))$ and its $\langle\cdot,\cdot\rangle$-orthogonal complement are both A-stable. It follows now by induction that n is even and that there is an orthonormal basis $\{v_1, v_2, \ldots, v_n\}$ of $\mathbb{C}^n$ with $b(v_1, v_2) = b(v_3, v_4) = \cdots = -1$ and all other $b(v_i, v_j) = 0$. Sending the v_i to the usual basis vectors of $\mathbb{C}^n$ transforms G into a subgroup of $Sp(n/2)$. (We note that $\mathbb{C}^n$ can now be interpreted as quaternion space $\mathbb{H}^{n/2}$, with A corresponding to the quaternion unit j, and that in this interpretation $Sp(n)$ consists of the $\mathbb{C}^n$-unitary quaternionic linear maps of $\mathbb{H}^{n/2}$ to itself.)$\surd$

From our earlier results we deduce with the help of Theorem H that all representations of Spin(n) for $n \equiv \pm 1$ or 0 mod 8, of $SO(n)$ for $n \equiv$ 2 mod 4, and of the compact groups G_2, F_4, E_8 can be transformed into real-orthogonal form.

We also note: The spin representation Δ_l of B_l is orthogonal for $l \equiv 0$ or 3 mod 4 and symplectic for $l \equiv 1$ or 2 mod 4; the half-spin representations $\Delta_l^{\pm}$ of D_l are orthogonal for $l \equiv 0$ mod 4, symplectic for $l \equiv 2$ mod 4, and not self-contragredient for odd l.

Appendix

Linear Algebra

The purpose of this appendix is to list some facts, conventions and notations of linear algebra in the form in which we like to use them. We follow pretty much the book by P. R. Halmos [9]. We use $\mathbb{R}$ (the real numbers) and $\mathbb{C}$ (the complex numbers) as scalars. Also, $\mathbb{N}$ stands for the natural numbers $\{1, 2, 3, \ldots\}$ and $\mathbb{Z}$ stands for the integers; finally $\mathbb{Z}/n$ or $\mathbb{Z}/n\mathbb{Z}$ stands for the cyclic group of order n, the integers mod the natural number n. We write $\mathbb{F}^n$ for the standard n-dimensional vector space over the field $\mathbb{F}$ (with $\mathbb{F} = \mathbb{R}$ or $= \mathbb{C}$ for us). Its elements are written as $(x_1, x_2, \ldots, x_n)$ with x_i in $\mathbb{F}$ and are considered as column vectors (occasionally the indices begin with 0). We denote by e_i the i-th standard coordinate vector $(0, \ldots, 0, 1, 0, \ldots, 0)$ with a 1 at the i-th place, and by ω_i the i-th coordinate function which assigns to each vector its i-th coordinate.

Vector spaces $(V, W, \ldots)$ are of finite dimension unless explicitly stated not to be so. For a subset M of a vector space V, we denote by $((M))$ the linear span of M in V. For a complex vector space V we write $V_{\mathbb{R}}$ for the real vector space obtained from V by restriction of scalars from $\mathbb{C}$ to $\mathbb{R}$; for a real vector space W we write $W_{\mathbb{C}}$ for the complex vector space obtained from W by extension of scalars from $\mathbb{R}$ to $\mathbb{C}$, i.e. the tensor product $W \otimes_{\mathbb{R}} \mathbb{C}$ (or, simpler, the space of all formal combinations $u + iv$ with u, v in W and the obvious linear operations).

An *operator* is a linear transformation of a vector space to itself. Trace and determinant of an operator A are written $\operatorname{tr} A$ and $\det A$. *The identity operator* is denoted by 1 or by *id.* Wr to a basis of the vector space an operator is represented by a matrix; similarly for linear transformations from one space to another. We write I for the identity matrix.

$\operatorname{diag}(\lambda_1, \lambda_2, \ldots, \lambda_n)$ stands for the $n \times n$ diagonal matrix with the λ_i on the diagonal; the λ_i could be (square) matrices.

The *dual* or *transposed* of a vector space V, the space of linear functions on V, is denoted by $V^{\top}$ (this deviates from the usual notation $'$ or *; "dual" being a functor, we like to indicate its effect on objects and morphisms by the same symbol). We note that $\{e_i\}$ and $\{\omega_i\}$ are *dual bases* of $\mathbb{F}^n$ and its dual.

For a linear transformation A from V to W (sending the vector v to $A(v) = A \cdot v = Av$) we write $A^\top$ for the *transposed* or *dual* of A, from $W^\top$ to $V^\top$ (defined by $A\mu(v) = \mu(Av)$ for μ in $W^\top$ and v in V). For invertible A the map $(A^{-1})^\top, = (A^\top)^{-1}$, is the *contragredient* of A, denoted by $A^\vee$. We use also a related notion, the *infinitesimal contragredient* $A^\triangle$ of any operator A (not assumed invertible), defined to be $-A^\top$.

The *adjoint* M^* of a matrix M is the transposed complex-conjugate.

As usual $\ker A$ denotes the *kernel* or *nullspace* of A (the set of vectors in V that are sent to 0 by A) and *im* A denotes the *image space* of V under A (the set of all Av, as v runs through V). There is the natural identification of V with its second dual $V^{\top\top}$ (this holds by our assumption on the dimension), which permits us to write $v(\mu) = \mu(v)$ for v in V and μ in $V^\top$, and to identify $A^{\top\top}$ with A. Composition of linear transformations is written $A \circ B$ or $A \cdot B$ or AB. Similarly matrix product is written $M \cdot N$ or MN.

Bilinear maps generally go from the Cartesian product $V \times W$ of two vector spaces V and W to a third space U. A *bilinear form* on V (denoted by $b, b(\cdot,\cdot), \langle\cdot,\cdot\rangle, \ldots$) is a bilinear map from $V \times V$ to the base field. Such a form defines two linear transformations from V to the dual $V^\top$ by the device of holding either the first or the second variable of the bilinear form fixed: to b we have $b' : V \to V^\top$ by $b'(u)(v) = b(u,v)$ and b'' by $b''(u)(v) = b(v,u)$; the two maps are transposes of each other(via $V = V^{\top\top}$. The form is symmetric iff the two maps are equal, and skew-symmetric, if they are negatives of each other. (We occasionally use "quadratic form" for "bilinear symmetric form"; that is permissible since our fields are not of characteristic 2.)

A *sesquilinear form* (for a V over $\mathbb{C}$) is a map from $V \times V$ to $\mathbb{C}$ that is linear in the second variable and conjugate-linear in the first variable.

A bilinear form b on V is *invariant* under an operator A if $b(Av, Aw) = b(v,w)$ for all v, w in V. We also use a related "infinitesimal" notion: b is *invariant in the infinitesimal sense* or *infinitesimally invariant* under A if $b(Av, w) + b(v, Aw) = 0$ for all v, w in V. (Cf.§1.3.)

A non-degenerate symmetric bilinear form, say b, is called an *inner product* or also a *metric* on V. One has then the canonical induced isomorphism (occasionally called the *Killing isomorphism*) $\rho \longleftrightarrow h_\rho$ between $V^\top$ and V, defined by $b(h_\rho, v) = \rho(v)$ for v in V and ρ in $V^\top$. Defining the form b on $V^\top$ by $b(\rho, \sigma) = b(h_\rho, h_\sigma)$ makes this isomorphism an isometry. An invertible operator A on V is an isometry precisely if it goes into its contragredient under this isomorphism. We use the terms inner product and metric also for *Hermitean forms* [i.e., sesquilinear forms with $\langle w, v\rangle$ equal to the conjugate $\langle v, w\rangle^-$ of $\langle v, w\rangle$], and occasionally for degenerate forms.

Let A be an operator on V. The *nilspace* of A consists of the vectors annulled (sent to 0) by some power of A. An *eigenvector* of A is a non-zero vector v with $Av = \eta v$ for some scalar η (the *eigenvalue* of A for

v). The *eigenspace* V_η, for a given scalar η, is the nullspace (NB, not the nilspace) of $A - \eta$ (i.e., of $A - \eta \cdot id$); this is the subspace 0 of V, if η is not eigenvalue of A. The *primary decomposition* theorem says that a complex V is direct sum of the nilspaces of the operators $A - \eta$ with η running over the eigenvalues of A (or over all of $\mathbb{C}$, if one wants). This is refined by the *Jordan form*: A can be written uniquely as $S+N$, where S is *semisimple* (= diagonizable), N is *nilpotent* (some power of N is 0), and S and N commute ($SN = NS$). The eigenvalues of S are those of A, including multiplicities (the *characteristic polynomial* $\chi_A(x) = \det(A - x \cdot id)$ equals that of S). Nilpotency is equivalent to the vanishing of all eigenvalues; in particular the trace is 0.

If a subspace U of V is *invariant* or *stable* under A (i.e., $A(U) \subset U$), then there is the induced operator A on U, and also on the *quotient space* V/U (= the space of cosets $v+U$), with $A(v+U) = Av+U$. The canonical *quotient map* $\pi : V \to V/U$ (sending v to $v + U$) satisfies $A \circ \pi = \pi \circ A$.

The last relation is a special case of *equivariance* : Let V and V' be two vector spaces. To each m in some set M let there be assigned an operator A_m on V and an operator A'_m on V' ("M operates on V and on V'"). A linear map $B : V \to V'$ is called *equivariant* (wr to the given actions of M) if $B \circ A_m = A'_m \circ B$ holds for all m. (One also says: B *intertwines* the two actions.)

A vector space with a given family of operators is called *simple* or *irreducible* (wr to the given operators) if there is no non-trivial (i.e., different from 0 and the whole space) subspace that is stable under all the operators.

A diagram $V' \to V \to V''$ of vector spaces, with maps A and B, is *exact*, if $\operatorname{im} A = \ker B$. A finite or infinite diagram $\cdots \to V_i \to V_{i+1} \to \cdots$ is exact, if each section of length 3 is exact. A *short exact sequence*, i.e. an exact sequence of the form $0 \to U \to V \to W \to 0$, means that U is identified with a subspace of V and that W is identified with V/U.

A *splitting* of a map $A : V \to W$ is a map $B : W \to V$ such that $A \circ B = id_W$. This is important in the case of short exact sequences where splitting either $U \to V$ or $V \to W$ amounts to representing V as direct sum of U and W. This is particularly important if one has an assignment of operators on V and W as above, and one tries to find an equivariant splitting of $V \to V/U$.

Let $m \to A_m$, for m in M, be an assignment of operators on V, as above. Let $0 = V_0 \subset V_1 \subset V_2 \subset \cdots \subset V_r = V$ be a strictly increasing sequence of subspaces of V, all stable under the A_m, and suppose that the sequence is maximal in the sense that no stable subspace can be interpolated anywhere in the sequence. Then each quotient is simple or irreducible under the A_m, and the Jordan-Hoelder Theorem says then that the collection $\{V_{i+1}/V_i\}$

of quotient spaces is uniquely determined up to order and equivariant isomorphisms.

All these notions apply in particular to the case that we have to do with frequently, where M is a group and where the assignment $m \to A_m$ is a representation of M, i.e., where the relation $A_{m \cdot m'} = A_m \circ A'_m$ holds for all m and m' in M.

A *cone* in a (real) vector space is a subset that is closed under addition and under multiplication by positive real numbers; cones are of course convex sets. A very special case is a *half-space*, a set of the form $\{v : \rho(v) \geq 0\}$, consisting of the points v where some non-zero linear function ρ takes non-negative values. The cones that we have to do with are finite intersections of half-spaces. Such a cone has for boundary (in the sense of convex sets, i.e. the usual point set boundary wr to the subspace spanned by the cone) a finite number of similar cones, of dimension one less than that of the cone itself, each lying in the nullspace of one of the defining linear functions. These faces are called the *walls* or faces of codimension 1. They in turn have faces etc., until one comes to the faces of dimension one, the edges, and the face of dimension 0, the vertex (the origin, 0).

More precisely, these are the *closed* cones. We will also have to do with *open* cones, the interiors of the closed ones; they can be introduced in a slightly different way, namely as the components of the complement of the union of a finite number of hyperplanes in the given vector space. (A *hyperplane* is the nullspace of a non-zero linear function.)

For two vector spaces V and W one has the *tensor product* $V \otimes W$ (similarly for more factors), and the associated concept of the tensor product $A \otimes B$ of two linear maps A and B. (Main fact: Bilinear maps $V \times W \to U$ correspond to linear maps $V \otimes W \to U$.)

There is also the notion of the *symmetric powers* $S^r V$ and the *exterior powers* $\bigwedge^r V$ of a vector space V (with the associated notion of symmetric power $S^r A$ and exterior power $\bigwedge^r A$ of a linear map A). We will treat them either as the usual quotient spaces of the r-th tensor power $V^{\otimes r}$ of V (i.e., $V^{\otimes r}$ modulo the tensors that contain some $v \otimes w - w \otimes v$, resp some $v \otimes v$ as factor) or as the spaces of all symmetric, resp skew-symmetric elements in $V^{\otimes r}$. For the standard properties of these constructions see, e.g., [17].

We note two general facts.

(1) *Schur's lemma* (which we will use often): Let $m \to A_m$ and $m \to A'_m$ be assignments of operators on vector spaces V and V', as above, and let $B : V \to V'$ be equivariant wr to these operators. Then, if V and V' are irreducible under the operators, B is either 0 or an isomorphism. In particular, if V is a complex vector space, irreducible under an assignment $m \to A_m$ of operators, and B is an operator on V, equivariant wr to the A_m, then B is a scalar operator, of the form $c \cdot id_V$ with some c in $\mathbb{C}$.

(2) In a vector space V with a (positive definite) inner product $\langle \cdot, \cdot \rangle$ we have the notion of adjoint A^* of an operator A, defined by $\langle Av, w \rangle = \langle v, A^* w \rangle$, and hence the notion of self-adjoint ($A = A^*$) and skew-adjoint ($A = -A^*$) operators. There is the *spectral theorem*: A self-adjoint operator has real eigenvalues, and the eigenvectors can be chosen to form an orthonormal basis for V.

(Note also the correspondence between self-adjoint operators and symmetric bilinear [resp Hermitean] forms in a real [resp complex] vector space, given by $a(u, v) = \langle Au, v \rangle$).

References

1. A.K. Bose and J. Patera, Classification of Finite-Dimensional Irreducible Representations of Connected Complex Semisimple Lie Groups, Journ. of Math. Physics **11** (1979), 2231–2234

2. R. Brauer, Sur la multiplication des caractéristiques des groupes continus et semisimples, C.R. Acad. des Sci. **204** (1937), 1784

3. É. Cartan, *Oeuvres complètes*, Gauthier-Villars, Paris 1952

4. H. Casimir and B.L. van der Waerden, Algebraischer Beweis der vollständigen Reduzibilität der Darstellungen halbeinfacher Liescher Gruppen, Math. Annalen **111** (1935), 1–12

5. C. Chevalley, Sur la classification des algèbres de Lie simples et de leur représentations, C.R. Acad. Sci. **227** (1948), 1136–1138

6. C. Chevalley, Sur certaines groupes simples, Tôhoku Math. J. (2) **7** (1955), 14–66

7. S. Coleman, *Aspects of Symmetry*, Cambridge U.Press, 1985

8. H. Freudenthal, Zur Berechnung der Charactere der Halbeinfachen Lieschen Gruppen, Indag. Math. **16** (1954), 369–376

9. P.R. Halmos, *Finite Dimensional Vector Spaces*, Princeton University Press, Princeton 1974

10. Harish-Chandra, On Some Applications of the Universal Enveloping Algebra of a Semi-simple Lie Algebra, Trans. Amer. Math. Soc. **70** (1951), 28–96

11. S. Helgason, *Differential Geometry and Symmetric Spaces*, Academic Press, New York 1962

12. S. Helgason, *Differential Geometry, Lie Groups and Symmetric Spaces*, Academic Press, New York 1978

13. N. Jacobson, *Lie Algebras,* J.Wiley & Sons, New York 1962

14. W. Killing, Die Zusammensetzung der stetigen endlichen Transformationsgruppen,I,II,III,IV, Math. Ann. **31** (1888), 252–290; **32** (1889), 1–48; **34** (1889), 57–122; **36** (1890), 161–189

15. A.U. Klimyk, Multiplicities of Weights of Representations and Multiplicities of Representations of Semisimple Lie Algebras, Doklady Akad. Nauk SSSR **177** (1967), 1001–1004

16. A. Korkin and G. Zolotarev, Sur les formes quadratiques, Math. Annalen **6** (1873), 366–389

17. B. Kostant, A Formula for the Multiplicity of a Weight, Trans. Amer. Math. Soc. **93** (1959), 53–73

18. O. Loos, *Symmetric Spaces I,II,* W.A.Benjamin, Inc, New York 1969

19. S. Lang, *Algebra,* Addison-Wesley 1969

20. I.A. Mal'cev, On Semisimple Subgroups of Lie Groups, Izv. Akad. Nauk SSSR, **8** (1944) 143–174; Amer. Math. Soc. Translations 33 (1950).

21. R. Richardson, Compact Real Forms of a Complex Semi-simple Lie Algebra, J. of Differential Geometry **2** (1968), 411–419

22. R. Steinberg, A General Clebsch-Gordan Theorem, Bull. Amer. Math. Soc. **67** (1961), 406–407

23. J. Tits, Sur les constants de structure et le théorème d'existence des algèbres de Lie semi-simples, IHES **31** (1966), 21–58

24. V.S. Varadarajan, *Lie Groups, Lie Algebras, and Their Representations,* Prentice-Hall, New York 1974

25. H. Weyl, Theorie der Darstellung kontinuierlicher halbeinfacher Gruppen durch lineare Transformationen, I, II, III, und Nachtrag, Math. Z. **23** (1925), 271–309; **24** (1926), 328–376, 377–395, 789–791

26. H. Weyl, *The Structure and Representation of Continuous Groups,* mimeographed notes, Institute for Advanced Study, Princeton 1933/34, 1934/35

27. H. Weyl, *The Classical Groups,* Princeton University Press, Princeton 1939, 1946

Index

Symbol Index

Universitext *(continued)*

Samelson: Notes on Lie Algebras
Smith: Power Series From a Computational Point of View
Smoryński: Self-Reference and Modal Logic
Stroock: An Introduction to the Theory of Large Deviations
Sunder: An Invitation to von Neumann Algebras
Tondeur: Foliations on Riemannian Manifolds
Verhulst: Nonlinear Differential Equations and Dynamical Systems
Zaanen: Continuity, Integration and Fourier Theory